How
Shoot, Edit and Distribute
HDV

The Complete, Up to Date Guide
to Working with the HDV Format

Andrew Lock

The Focal Point Publishing Company
Orange, CA

How to Shoot, Edit and Distribute HDV
Copyright 2006 by Andrew Lock
All rights reserved

ISBN: 0-9777441-0-8
Printed in the U.S.A.

No part of this publication may be reproduced or distributed in any form or by any means, or stored in a database or retrieval system, without the prior written permission of the publisher. Exceptions are made for brief excerpts used in published reviews.

The Focal Point Publishing Company

This book is designed to provide accurate and authoritative information with regard to the subject matter covered. It is sold with the understanding that the publisher and/or author are not engaged in rendering legal, accounting, or other professional advice. If legal advice or expert assistance is required, the services of a competent professional person should be sought directly.

Neither Andrew Lock nor The Focal Point Publishing Company can be held responsible for loss or injury as a result of any unsafe behaviour by any reader who undertakes to recreate any action described in this publication. Using the information contained within the text of this book is done solely at the readers own risk.

Cover images are courtesy of Sony Electronics Inc, JVC, and Canon. All rights reserved.
HDV and the HDV logo are trademarks of Sony Corporation
and Victor Company of Japan, Limited.

Registered trademarks and trademarks are the property of the
registered trademark and trademark owners.

This book contains trademarks and registered trade names of multiple companies. Appearance of such trade names and trademarks is not intended to imply endorsement or otherwise by The Focal Point Publishing Company or Andrew Lock.

What Others Say About This Book

"Here's a real-world look at the sometimes contradictory world of HDV, written by someone who knows what he's talking about. Sidestepping any snobbery and market-speak, this book will painlessly introduce you to the new world of HDV production without getting bogged down in needless technical details."

- *Charlie White, Emmy Award Winning Producer & Technology Journalist*

"Complete clarity, amazing detail, just like HDV cameras....and completely readable."

- *James MacGregor, Editor— 'Wideshot'*

"In an age where there are an endless supply of questions regarding formats, functions and practicality it is difficult to get concrete answers. Andrew Lock's book on HDV gives you answers about the latest HDV cameras as well as covering the essentials you need to know in order be more efficient with this new format."

- *Jacob Rosenberg—Filmmaker and Author*

"High Definition is the future and HDV is how most of us will get there. Andrew Lock makes this terrific format accessible for beginners and essential for the experts. If you own an HDV camera, this book will make that marvellous tool come alive. If not, buy this book and go shopping for your camera."

- *Greg Lefevre, Former CNN Bureau Chief, www.greglefevre.com*

"This book is a must for anyone contemplating the jump into HDV shooting and post-production; clearing away the misinformation with a concise and accurate guide to the exciting world of inexpensive HD production."

- *David Newman, Chief Technology Officer, CineForm*

"HDV...Now I get it! This book is a must read for anyone planning to stay in the video production business."

- *John Goolsby, Author of "The Business of Wedding & Special Event Videography"*

"By explaining the format in lay terms, describing the cameras that shoot it, giving tips on where to buy gear and reviewing the best editing solutions, ...Andrew's book puts a practical face on the technical side of HDV."

- *Dominic Milano, Editor in Chief, DV Magazine*

Clear, concise and easy to comprehend, Andrew Lock's new book really takes the mystery out of HDV. If you're considering HDV and need to know more about it, this excellent book is a must-read."

- *Steve Yankee, Founder: www.VideoBusinessAdvisor.com*

Acknowledgements

Over the years it's been my pleasure to get to know many interesting people from all parts of the globe. As a result, I have forged numerous friendships, and the support and interest of these friends was a large factor in spurring me on to write this book. For that, I am deeply grateful, thank you.

It's only appropriate to express my appreciation to David Newman from CineForm who acted as technical editor for this book. If you've been around any of the online forums on HD, you'll know that David gives very generously of his extensive knowledge and experience, and he is well regarded by all. The accuracy of the technical information in this book is in no small part due to David's willingness and diligence in reviewing the material.

There are inevitably some errors in this book, grammatical or otherwise. If you find any that trouble you, please email us at: focalpointpub@gmail.com and we'll rectify them in the next edition.

Special thanks go to my family who are unwavering in their encouragement, and particularly my wife Luci, who constantly puts up with my unconventional (but successful) approach to business, and who has never failed to offer her love and support for all my projects.

This book is available at special quantity discounts to use as a training aid in schools, colleges and other academic institutions. For more information, please email: focalpointpub@gmail.com

The author, Andrew Lock, is available for a limited number of speaking engagements and consulting assignments. For information, email: Andrew Lock at:mail@andrewlock.com

Contents

Chapter 1	The Road to HD…
Chapter 2	HD and HDV Defined
Chapter 3	Getting Technical
Chapter 4	Overview of HDV Cameras
Chapter 5	JVC GR-HD1
Chapter 6	JVC GR-HD10
Chapter 7	Sony HDR-FX1
Chapter 8	Sony HVR-Z1
Chapter 9	JVC GY-HD100
Chapter 10	Sony HVR-A1
Chapter 11	Sony HDR-HC1
Chapter 12	Canon XL-H1
Chapter 13	Buying Advice
Chapter 14	How to Shoot HDV
Chapter 15	Media and More
Chapter 16	How to Get the Film Look
Chapter 17	How to Edit HDV

Chapter 18	Editing Hardware and Software
Chapter 19	Choosing and Using Computers for HDV
Chapter 20	Useful HDV Devices & Accessories
Chapter 21	Monitoring HDV
Chapter 22	How to Distribute HDV
Chapter 23	Conclusion
Resource	HDV Related Websites and Resources

Chapter 1
The Road to HD...

Do you remember the first camcorder you used or owned? Maybe it was one of those hefty cameras attached by a chunky cable to a separate 'portable' VHS recorder, or perhaps it was one of the later (still unwieldy) VHS or Video8 camcorders. At the time we all thought they were wonderful. We eagerly gathered our family around the TV, marvelling over how amazing it was to watch video that we had recorded ourselves.

We've come a long way from cameras like this

Let's continue the history lesson...

In 1995, DV (Digital Video) made a grand entrance with the introduction of the Sony VX1000. Wow, what a difference. Suddenly the picture quality was much sharper and more vibrant, a huge improvement. With further milestone cameras such as the Sony VX2000, both consumers and semi-professionals were suddenly able to record video that seemed to rival the quality broadcast by the national TV networks.

What made Digital Video so much better than analog video? Well, even if you used a good quality lens, analog video easily introduced errors, tape noise and interference which degraded the picture. Then the moment you tried to edit the footage, or even make a simple transfer from one tape to another, the picture was degraded further. Digital video overcame all these problems.

The Road to HD

By the late 1990's the technology had reached a point where it seemed unlikely that any improvements could be made to surpass the quality of DV. After all, even some TV broadcasters and production companies were (and still are) routinely using cameras such as the Sony VX2100 (or PD170). These DV camcorders produce excellent quality video at a price point that would have been unthinkable ten years ago.

With the quality gap between consumer and professional video equipment dramatically narrowed, could there really be any further advances?

Of course, the answer lies with HD (High Definition). HD video is to DV what DV was to analog video. In fact, it's no exaggeration to say that HD is the most important advance in video since we progressed from black and white images to color television. Think about it—while DV made noticeable improvements over analog video, HD has made a quantum leap in both picture quality and sound. It has already revolutionised TV production in the USA to the extent that most major sitcoms, dramas and other popular shows are now recorded in HD.

Admittedly it's still going to take a little while for the entire population to move over to the new format, as they will need to buy an HD capable TV, as well as subscribe to channels that transmit HD programming. The fact remains, HD is here to stay, and it is already having a huge impact on consumers.

How widespread is the current use of HD? Most major TV production companies in the USA now produce much of their programming in HD. The BBC has stated that they plan to produce all of their programming in HD by 2010, and Hollywood studios are positively salivating at the prospect of re-releasing their back catalogues of movies in HD format on the new HD-DVD or Blu-Ray discs.

How to Shoot, Edit and Distribute HDV

George Lucas—the famous creator of the *Star Wars* movies—reportedly stated that he is so pleased with the workflow and results of HD that he will never shoot another movie on film stock again. Actually, we all owe a debt of thanks to Lucas because he has blazed a trail in creating HD movies—not only did he work through the learning curve of a new format, devising solutions and introducing technology enhancements in the process, he was also subjected to much criticism from poorly informed movie industry purists who were adamant that video could never compare to the look of film.

Amazingly, it's conceivable that within the next ten years we might reach a point where more new movies are shot using HD video cameras than with film.

It's an exciting time ahead, and HDV is adding to the excitement. Read on to find out why...

The picture quality of High Definition on a properly configured HD TV is often described as though one is "looking through a window"

Chapter 2
HD and HDV Defined

Just how good is the quality of HD? Well, most people literally drop their jaw when they first see well shot HD images on a large, properly configured, high resolution plasma screen. Frequently you will hear comments such as, "It's like looking through a window" or "I feel as though I can reach out and touch the objects on the screen."

Such comments might sound exaggerated, but in truth, never before have we been able to view such lifelike TV images within our homes, on screens that are larger, flatter and thinner than ever.

Let's not forget about the sound quality too. Within HD is the capability of playing fully digital surround sound, much like you hear in a modern movie theater. Remember, sound makes up a surprisingly large percentage of the enjoyment of any program, yet digital audio is a key feature of the HD format that is often overlooked.

So what exactly is HDV? How is it different from HD? Well, let's start by explaining what HD is. 'HD' is an abbreviation for 'High Definition' and it's used as a general term that encompasses *any video that is higher quality than standard definition video* (as measured by the pixel size of a single frame). As you'll come to see, contrary to popular belief, HD is not a single format, it actually encompasses a range of frame sizes and rates.

To help you understand the quality of HD as it fits into the image 'hierarchy,' if VHS is at the bottom of the ladder, and IMAX film is at the top, HD sits between 16mm and 35mm film, so it has a higher resolution than 16mm film, but not as high as 35mm or IMAX.

Note that prior to HD, no video format came remotely close to the quality of film, so you can see what a significant advancement it is. Incidentally, looking at it from another point of view, that's the reason why many movies can be seen on HD TV channels in such great quality—there is enough image information (quality) in a 35mm movie film to down-convert it to HD.

HDV is Born

In July 2003, a consortium of manufacturers made up of Sony, JVC, Canon, and Sharp, announced the specification for a new format to be known as HDV. HDV stands for '*High Definition Video*' which might lead one to think that HDV is the same as HD? It certainly sounds the same, doesn't it?

SONY

JVC

Canon

SHARP

The Consortium of Manufacturers that initially agreed on the HDV format

If you refer back to the definition of HD, HDV complies with that definition, so **HDV can rightly be called High Definition**. However, HDV was also given a dedicated, specific name to be identified as a separate format in its own right. So logically there must be some differences, otherwise it would also be called HD!

HD and HDV Defined

The answer lies in the fact that there are many 'flavors' of HD, so HDV is actually a subset of HD. Let's use an analogy to explain this further. Imagine that in your town there was only one steakhouse, a high-end restaurant that had the best cuts of steak served in opulent surroundings with a price to match. However, eventually an inexpensive steak house appeared in your town, offering steak dinners at a substantially cheaper price. At both restaurants you would be eating steak, but clearly the inexpensive steak house has to cut a few corners in order to offer you a steak meal for a greatly reduced price.

HDV is the video equivalent of the inexpensive steak dinner. It is still HD, the premium quality format, but it has cut some corners and made some compromises in order to work within a much lower budget.

During the early years of HD production, there was only one option for capturing HD footage, you had no choice but to work with very expensive high-end equipment—HD cameras that cost more than $65,000 for example. HD was definitely out of the price range that consumers could afford, and even beyond the reach of most small to medium production companies. The words 'inexpensive' and 'HD' were simply never seen together!

In essence then, *HDV was created as an inexpensive way to capture and work with High Definition video.* Although high-end HD equipment still commands high prices and produces the ultimate picture quality, HDV has paved the way for low-cost HD production. As the initials of the format imply, HDV is a combination of HD and DV—it is HD recorded on to DV tape.

HDV is HD on a budget. It's also been called HD for the masses because it has opened up the opportunity for many more people to work with High Definition. Keeping things in perspective,

How to Shoot, Edit and Distribute HDV

HDV is not yet as cheap as DV, and definitely not as easy to edit, so it has its own challenges. There are 'costs' from compromise.

High Definition on a Budget—How?

You may be wondering how it was possible to invent an inexpensive way of working with High Definition. Well, there were two main factors that enabled this development.

Firstly, manufacturers realized that they would need to use an existing tape format, and Mini-DV was the logical choice as it forms the basis for the vast majority of camcorders on the market today. By using the existing Mini-DV tape size along with the associated manufacturing processes and hardware, manufacturers were able to substantially reduce the costs of both inventing the HDV format and producing HDV hardware.

Secondly, since it was important to be able to record at least one hour of HDV footage onto a Mini-DV tape, manufacturers realized that the new HDV camcorders would need to use advanced compression techniques to 'squeeze' all the additional data onto a Mini-DV tape.

The solution that was devised is a variation of MPEG-2 compression. Interestingly, although you would expect the picture to be noticeably lower in quality compared with uncompressed HD, the compression algorithm used in HDV camcorders is very sophisticated and as a result the picture quality is superb. Of course, with more expensive monitors you can definitely see the difference in quality between HD and HDV, so most broadcasters won't be getting rid of their high-end, expensive HD cameras anytime soon.

Let's say a few words about the 'snobbery' factor involved in all of this. If you had paid out close to a hundred thousand dollars

HD and HDV Defined

for an HD camera, what would you think about someone who comes along and claims that their $5000 camera can do the same thing? Naturally you would vigorously defend your purchase, and that's exactly what is happening in the current scene. Many users of high-end HD equipment decry HDV saying it's "not broadcast quality", "not HD", "solely a consumer format"...you get the idea.

As we've established, HDV *is* considered High Definition according to the official definition of HD. The quality is stunning for the price point, and yes it is broadcast quality. Is it as good as the high-end HD cameras? No, definitely not, nonetheless HDV is being use in the broadcast world.

You may be interested to know that it was reported in the trade press that the BBC's technical engineers closely examined the quality of Sony's Z1 HDV camera when it was released, and concluded that it is more than adequate for use in TV broadcast. In fact the same report cited that they also consider it to be the natural replacement for the Sony PD170 for standard definition video acquisition. Furthermore, it's been well publicized that many TV crews around the world are already using the Z1 for on location news gathering, documentaries, reality shows and more.

In answer to the pointless debate that continues, HD and HDV both serve useful functions and there is room for both.

One thing is clear. As noted earlier, HDV is a significant breakthrough, a milestone in video technology. The ability to record High Definition video at a price point that is within the reach of some consumers would have been unthinkable even a few short years ago. With some fifty-five companies currently backing HDV, it's inevitable that we will now see a rapid acceleration in the support of the format as well an increase in the quantity of High Definition content available to view on TV.

Chapter 3
Getting Technical

It's not the intention of this book to get bogged down with unnecessary technical details about video engineering. There are plenty of other books out there that cover the technical details in great depth if that is of interest to you.

This book is definitely for the non-engineer. Its aim is to provide you with tips, techniques, and practical guidance to help you.

Having laid that framework, there are a few technical aspects of HDV that you should be aware of, because an understanding of them will make it easier to transition to an HDV workflow.

As tempting as it may be, don't skip over this chapter. It's been simplified as much as possible, and your efforts to review the information presented will be rewarded.

Frame Sizes

The main implication of working with HDV is that the frame size is substantially larger than with standard definition DV. Let's discuss frame sizes so you can see the significance of how these relate to both shooting and editing with the HDV format.

Firstly, HD or HDV is *always* recorded in the widescreen, 16:9 aspect ratio, DV by comparison can be either 4:3 or 16:9. We'll talk more about the differences of shooting in widescreen in the chapter on shooting techniques, but for now just remember that HDV is *widescreen only*.

If you take a look at the chart on page 12, you will see that there are numerous 'flavors' of HDV. At the time of writing each

Getting Technical

manufacturer has indicated a preference for a specific variation of HDV. These two variations are named HD1 (720p) and HD2 (1080i) according to the official specifications of the format—not HDV1 and HDV2 as they are often wrongly referred to by some.

There are two key combinations of measurements that make up these different variations of HDV—*frame size* and *frame rate*.

Frame size, as the name suggests is defined as the physical dimensions of each frame of video, as measured in pixels. The standard way of indicating this is:

Number of pixels wide x Number of pixels high

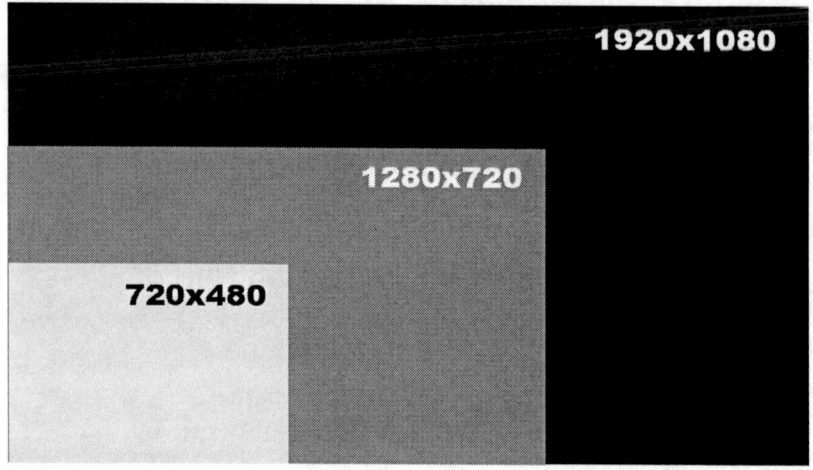

The relative frame sizes of standard definition (720x480) and HD (1280x720 & 1920x180)

The second measurement is 'frame rate over time', which is measured in frames per second. This is simply the number of frames that are displayed each second. When we watch video it is easy to forget that what we perceive as motion is simply a continuous stream of individual static images, these are called 'frames.' For example, when you press the freeze frame button on your remote, it holds or freezes the picture on screen—what you see there is just one frame.

To further grasp this concept, think about the way that an animated movie is produced. Figures in a scene are setup either as 'live action' miniatures, or are hand drawn by an animator in their starting positions. The camera captures a single frame, then the figures are moved a fraction or redrawn to indicate a slight movement, and the next frame is captured. That process is repeated over and over again, painstakingly advancing the movement *frame by frame*.

When the film is played back, the succession of rapidly changing frames gives the illusion that the character is moving smoothly. Hence, the number of frames that appear each second is known as the frame rate.

That's the basic description of frame rate. However, with video there is a slight complication of this concept which we will cover after you've looked at the chart in more detail.

Now that you are familiar with the terms 'frame rate' and 'frame size', let's discuss how they relate to HDV. For one thing, the highest specified frame size within the HDV format, a 1440x1080 HDV frame, has about four and a half times as many pixels as a standard definition TV frame. Four and a half times the pixels! That's an enormous increase, but it's still less than the highest specification of an HD frame, which runs at 1920 x 1080 pixels, and contains approximately six times more data (pixels) than DV.

We've established then that the volume of pixel data involved with HDV is far greater than with DV.

Something else to keep in mind is that the largest frame size of HDV, 1440 x 1080, is stretched on output with a pixel aspect ratio

Getting Technical

of 1.33, to produce a full resolution HD frame of 1920 x 1080. This is because 1440 is the practical limit of horizontal pixels that can be recorded within the constraints of current HDV equipment.

Format	Pixel Resolution of one frame (Width x Height)	Total Number of pixels	Aspect Ratio (Width x Height)	Frames per second
NTSC DV	720 x 480	345,600	4:3	30 (60i)
PAL DV	720 x 576	414,720	4:3	25 (50i)
HDV	1280 x 720	921,600	16:9	24p
HDV	1280 x 720	921,600	16:9	25p
HDV	1280 x 720	921,600	16:9	30p
HDV	1280 x 720	921,600	16:9	50p
HDV	1280 x 720	921,600	16:9	60p
HDV	1440 x 1080	1,555,200	16:9	24p
HDV	1440 x 1080	1,555,200	16:9	25p
HDV	1440 x 1080	1,555,200	16:9	30p
HDV	1440 x 1080	1,555,200	16:9	50i
HDV	1440 x 1080	1,555,200	16:9	60i
HD	1920x1080	2,073,600	16:9	60i

Chart: DV, HDV and HD Technical Specs Comparison

Please note that in the chart, the frames per second figures for NTSC displays have been rounded up to the nearest number. Strictly speaking, as drop frame figures they should be indicated as 23.98 for 24, 29.97 for 30, and 59.94 for 60.

Note also in the chart, the 'p' and 'i' designation after the frame rate (frames per second) number. In this context, 'p' stands for progressive; 'i' stands for interlaced.

Progressive means that each frame is unique and contains the complete image, whereas each interlaced image is in fact half a frame.

How to Shoot, Edit and Distribute HDV

With interlaced video, each complete frame is split into two 'fields'—one field displays the odd lines in a frame, followed quickly by the other field which displays the even lines of the same frame. The two fields are interlaced (joined together) to form a complete frame.

> Don't get bogged down with trying to memorize all the numbers in this chapter. They are incidental and included purely to emphasize how much more pixel data is included in HDV compared with DV. It is helpful, but not essential to have an understanding of some of the technical background to HDV.

Something else to note on the chart are the 'frames per second' numbers for each 'flavor' of video. As you probably know, standard definition NTSC DV (the TV standard in the U.S. and some other countries) uses 30 frames per second (also known as 60i) whereas PAL DV (used in the UK and much of Europe) uses 25fps (50i). Similar 'frames per second' modes (25 & 30) are available for HDV, with some additional options—24p, 50p, and 60p.

Interestingly, with HDV the differences between PAL and NTSC recording formats have been narrowed to the extent that officially HDV does not make reference to PAL and NTSC anymore. The resolution and color space are common across all HDV frame rates, so when we're talking about the different flavors of HDV, it's more appropriate to refer to the different frame rates than to use the terms PAL or NTSC, (which are specific formats of broadcast video).

24p Video is Finally Here

Since 35mm movie film cameras shoot at 24 frames per second

Getting Technical

progressive (each frame has the full image), many video producers (and film makers for that matter) have been eagerly awaiting camcorders that record video at the same frame rate as film. They like the look of 24 frames per second as it obviously bears a closer resemblance to film, and they also like the fact that video shot at 24 frames per second (23.98 to be exact) can be more easily transferred to film for projection in a movie theater.

> **The 24p Trend**
>
> Since most modern TV display technologies such as Plasma, LCD, and DLP projection are now optimized for progressive rather than interlaced video, it makes sense for manufacturers to offer equipment that captures video at 24fps. That way, the entire workflow from camera to screen can stay in 24fps. It will be interesting to see how this trend develops over the next five to ten years when we reach a point where the majority of the population own TV displays that are optimized for progressive content.

It's easy to see why 24, 25, & 30 fps options are specified with HDV, but why the 50p and 60p 'frames per second' options? Quite simply, more frames means smoother motion. Admittedly, not everyone prefers the look of the faster frame rates, but it's interesting to have the option with this new technology.

What quantity of data actually records to tape? As noted previously, HDV uses clever MPEG-2 compression to enable a Mini-DV tape to store the substantial amount of data that comprises High Definition video, while maintaining a standard one hour recording time of Mini-DV. This is an important aspect to understand, because the storage of HDV is an area that is commonly misunderstood. There is a lot of misinformation about HDV out there. For example, many believe that HDV streams contain a huge amount of data, but because MPEG-2 compression is used, that's not the case. The amount of actual data on the tape is comparable to DV.

HD1 (720p) produces a stream of data at about 19Mbps (Megabits per second) and HD2 (1080i) runs at 25Mbps, which is the same as DV. This means that a single, 7200RPM hard disk can capture HDV with ease. Playback and editing is a different story, but more on that later. Suffice to say at this point that the complication with HDV is not with the data rate that the camera puts out, rather it's the *processing power required to decode the MPEG-2 compression.*

Something else to note about the HDV format is that it uses a 4:2:0 colorspace, compared with the 4:1:1 of NTSC DV. Without getting into complex details about colorspace, the color information in HDV makes it more robust than DV, so that's an area of improvement.

HDV Audio

In order to use the Mini-DV tape format, it wasn't just the video that needed to be compressed. After the video compression, there is little room on the tape for the audio, so that also needed to be compressed. The compression used is MPEG1, layer II at a data rate of 384kbps. It's not as high quality as DV audio, and it's not CD quality, but audibly it's as near to CD as to make it virtually indistinguishable. Considering the fairly aggressive compression ratio of 4:1, there has been much praise for the (relative) quality of audio within the HDV format.

If quality audio is a prime consideration of a shoot, for example if you were recording a concert, it would be sensible to record the audio separately to a minidisk, DAT or other digital recorder. That advice applies regardless of the recording format you are using. It's always preferable to record audio separately to retain the highest quality.

Chapter 4
HDV Cameras Overview

As much as many people have tried to 'pigeon hole' HDV, it is a little difficult to classify it in the grand scheme of all things video. You see, officially HDV was dubbed as a replacement for DV within the consumer realm of digital video. The reality is that like DV, there are a wide range of cameras on offer—some are clearly geared towards consumers whereas others are distinctly for professionals.

What makes the designation of HDV even more difficult is that all HDV cameras produce stunning picture quality, so even camcorders designed for consumers could be used effectively by knowledgeable professionals. So HDV crosses the boundaries of both consumer and professional worlds.

At the time of writing, there are eight HDV format camcorders available: three from JVC, four from Sony and one from Canon.

As a side note, Panasonic has released an HD camera, the AJ-HVX200, which records 100Mbps DVCPro HD directly on to P2 storage media cards (removable flash memory), or DV onto Mini-DV tapes. Although it's not an HDV camcorder, it's worth making a mental note of this camera as it records both 1080 (interlaced and progressive), and 720p, and it fits into the price range of HDV camcorders, which in itself is an incredible achievement. Panasonic make superb equipment, and this camera is no exception, but the problem right now is that the storage cards are relatively low capacity and incredibly expensive. Until those factors change, this camera will probably be a tough sell.

JVC were the first to launch an HDV camcorder, and with each successive release since then, the various manufacturers involved

How to Shoot, Edit and Distribute HDV

with the format have continued to raise the bar dramatically in terms of features. Here is a list of HDV camcorders, along with a remark about how the respective manufacturers positioned the camera in the marketplace (for consumer or professional use):

HDV CAMCORDERS LINE-UP

JVC GR-HD1 (Consumer)

JVC JY-HD10 (Pro)

JVC JY-HD100 (Pro)

SONY HDR-FX1 (Consumer)

SONY HVR-Z1 (Pro)

SONY HVR-A1 (Pro)

SONY HDR-HC1 (Consumer)

CANON XL H1 (Pro)

HDV Cameras Overview

The debate amongst manufacturers as to whether HD1 (720p) or HD2 (1080i) is better rages on as they each promote their cameras.

When you take all the marketing jargon out of the equation, it's fairly evident that each format is better for different purposes. For example, in general terms the 1080i format is more suited to viewing both slow motion and high speed motion, whereas 720p in general is a better fit for projects where you need to grab still frames, or analyse footage frame by frame. Likewise, 720p at 24 frames per second is clearly a better fit for projects that are destined for film (or to obtain the film look).

There are other pros and cons that are discussed within the reviews of each of the camcorders.

It's also worth noting that although manufacturers have positioned each camera for a specific market, there are clearly certain cameras that could span across both consumer and professional markets. The Sony FX1 is one such example. Officially it sits at the top of Sony's consumer range, but many professionals are using it effectively.

In terms of price, the lowest priced HDV camcorder is the Sony HC1, and the most expensive is the Canon XL H1. The maximum price threshold for manufacturers producing HDV cameras seems to be around $10,000, although that will probably change over time.

In the chapters that follow, each HDV camera is reviewed in order of its release, to give you an idea of how the technology has already improved in this relatively short space of time.

Chapter 5
JVC GR-HD1

Released in 2003, this was the first camcorder to use the HDV format. It was launched as part of JVC's consumer range of camcorders and it was a brave move for JVC. Unfortunately, because there was no support from software providers, it was dismissed by many as something of an oddity. That criticism was unfair because at the time it was a groundbreaking camera that produced excellent images under the right conditions.

Image courtesy of JVC

The camcorder features an HD resolution of 1280x720 @ 30p, as well as two other modes—DV and an 'SD' mode. The DV mode records in 4:3 or 16:9 aspect ratio as per most other DV camcorders, whereas the SD mode records 16:9 at 60p, using a similar MPEG-2 compression as the HDV mode. As an interim option, the real world practical use of the "SD" mode is debatable, other than perhaps for slow motion shots in SD productions (sixty progressive frames per second produces excellent slow-motion).

Poor Low-Light Performance

Unfortunately the single, 1/3" CCD (1.18 megapixels) is what lets the camera down when compared to more recent HDV cameras. Colors are not as rich as the newer HDV camcorders, and its low lighting capabilities leave a lot to be desired. Having said that, with good lighting it produces a decent image, and considering its relatively low price point of around $3500 on release, it truly opened up the possibilities for recording High Definition footage.

JVC GR-HD1

Nowadays, some years down the road, the camera can be purchased used, for under $2000.

Other features of the unit are the ability to capture still photos at 1280x720 resolution (significantly higher resolution than the still capture capability of standard definition camcorders), and MPEG-4 video, both to a removable 'SD' storage card.

The lens is a 52mm diameter, 10x optical zoom unit, with a focal range of 5.2mm to 52mm, and a maximum aperture of f/1.8. It is a fixed lens (non removable). The lens is very high quality for a camera in its price range, in fact the significant additional detail present in HDV *requires* high quality optics. It's reassuring to know that manufacturers can't skimp on lens quality with HDV camcorders.

Optical image stabilisation is a feature increasingly being used on semi-pro cameras and the OIS on this JVC unit works well. The level of OIS (sensitivity) is selectable via the menu, an option seen in few cameras prior to this one.

Optical stabilisation rather than digital means there is little degradation in the image quality. Digital stabilisation on the other hand, like digital zoom, lowers the picture quality because the pixels already present in the image are processed.

The weight of the camcorder with a Mini-DV tape and a battery is approximately 3.4 pounds.

Unusually for a consumer grade camcorder, the HD1 has numerous manual controls for functions such as iris, white balance, shutter speeds and focus. Sadly however, manual control of the audio channels is not provided. The only redeeming factor is that the AGC (automatic gain control) can be turned off if necessary in order to prevent the annoying sharp ramping of the audio levels during the recording of quiet scenes.

Audio input is via a standard mini jack socket, which is normal for a consumer grade camcorder.

The live and recorded images can be monitored by a color LCD screen, set in the traditional location as a flip-out unit on the left side of the camcorder. The LCD measures 3.5", with 113,000 pixels. Surprisingly it is configured in the 4:3 aspect ratio. Considering the camcorder is primarily designed to record HDV, it seems odd that the monitor would be supplied in the 4:3 aspect ration rather than 16:9. As a result, when shooting 16:9, the image appears in a letterbox format on the 4:3 screen. Very odd.

Playback capabilities are standard DV, the special SD mode mentioned earlier, 720p, and 1080i. Additionally the camcorder has the ability to both up-convert and down-convert to and from these formats which is very useful. Obviously the up-convert feature is limited to the resolution of the existing recorded data, in other words it does not result in the same quality of HD image that you would get from recording in native HDV mode.

Output connectors are composite (the single yellow RCA connector), S-Video (4 pin Y/C connector), component, and Firewire (for both input and output). The component connectors can only output recorded HDV, whereas the other connectors can output all recorded formats.

Many users of the camera were critical of the over-enthusiastic edge sharpening that the camera applies to the image. This was especially the case for those who were trying to obtain a film look, because the sharpening made the image look even more like video. This issue was changed on the later model, the HD10, but ultimately it is still a matter of personal preference.

Chapter 6
JVC JY-HD10

With an original retail price of around $4000 on release, JVC's HD10 camcorder is very similar to the HD1. However, it was released with additional features to serve the needs of the semi-pro or pro film-maker.

The most obvious difference externally on this camcorder is the inclusion of XLR audio connectors to accommodate professional microphones. Also, the microphone levels are shown in the viewfinder/monitor, but there are still no manual controls to adjust the levels, which is very surprising and disappointing for a 'professional' camera.

Image courtesy of JVC

Although not immediately apparent visually, the flip-out LCD screen has a higher resolution screen of 200,000 pixels (compared with the HD1).

How to Shoot, Edit and Distribute HDV

If you recall the earlier comments about having too much edge sharpening on the HD1, apparently JVC listened to users opinions, because the marketing materials for the HD10 comment that JVC's engineers tweaked the internal sharpness settings of this unit, with an edge enhancement processing that results in a generally better looking image. Ultimately, this 'enhancement' is still a matter of personal taste. Users have hotly debated over which image looked more natural—the HD1 or the HD10. It's obviously an area of personal preference.

Another major addition on the HD10 is an internal color bar generator, which aids in the setup and calibration of the camcorder on an external monitor.

On the negative side, just like the HD1, the HD10 is let down by its poor low light performance. If you are used to using a camera such as the Sony VX1000 or VX2000 you would be shocked at how the HD1 and HD10 struggle in low light.

When there were no other HDV camcorders to compare, these JVC cameras offered amazing resolution and quality for an equally amazingly low price. Admittedly, both the HD10 and the HD1 cameras are now starting to show their age when compared with more recent offerings, but JVC deserves a lot of credit for their bold move as the first player into the new HDV market.

In view of the above, this camera might be a good idea for someone who wants to dip their toe into the world of HDV without a huge expense. With good lighting conditions, the camera is still worthy of consideration if you are on a tight budget.

Chapter 7
Sony HDR-FX1

While JVC achieved a significant head start in the HDV race, Sony used the interim period wisely to learn a lot about the technology as well as carefully research what the market wanted. As a result, when they introduced their first HDV camcorder, the HDR-FX1, in November 2004, it proved to be a milestone camera in the world of video. In years to come, the FX1 will long be remembered as the camcorder that put HDV firmly on the map.

Although officially designated as a consumer camera, and offered for sale in 'Sony Style' stores as their top of the range camcorder, it appears to be more popular with semi-professionals and even professionals than wealthy consumers.

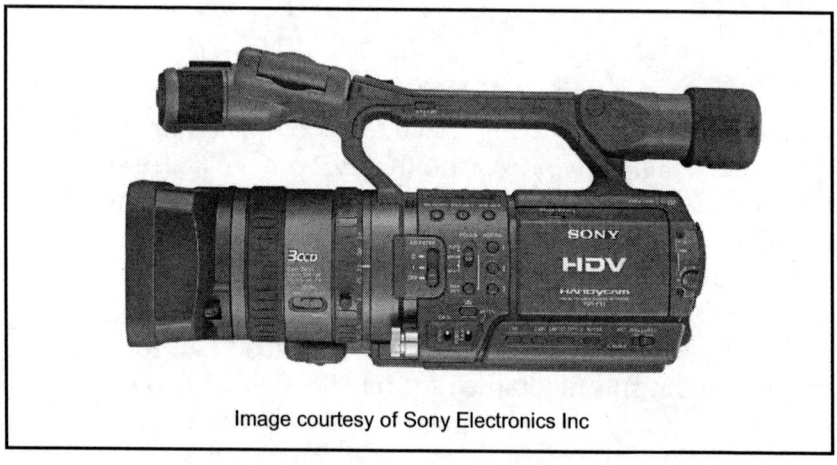

Image courtesy of Sony Electronics Inc

Weighing in at 4lbs 4oz, the FX1 is heavier than any of its DV counterparts, and boasts many differences and improvements compared with the earlier JVC cameras. Firstly, the ability to record 1080i onto Mini-DV tapes is truly groundbreaking. This of course is a much higher resolution than the 720p that the JVC cameras record. Remember, there are a variety of flavors of HDV formats, and for the time being JVC has committed to 720p (HD1) whereas Sony has chosen to go with 1080i (HD2).

The ability to record Standard DV is included, with a choice of 4:3 or 16:9 (widescreen) modes. Most users will agree that the standard definition recording modes are easily comparable in image quality to the Sony VX2000 or VX2100. That's quite a significant statement to make, because when you consider that you can purchase the FX1 for little more than the SD cameras just mentioned, *and gain the benefit of HDV recording* in the process, it makes it a very attractive camera to purchase. You are effectively getting two cameras in one.

Incidentally, this camera is available in two variations—one for North American markets and another for the UK and most of Europe. The U.S. model records in 1080 (60i) whereas the European version records 1080 (50i). Both versions feature real-time down-conversion of HDV to DV. This is a very useful and powerful feature because it means you can record HDV footage to tape and retain the option to play that footage back later (and capture it to your computer via Firewire) *as either High Definition or standard definition DV*—you get the best of both worlds.

How might this feature be used in a real world setting? Well, you could shoot a project in HDV mode with the intention of producing an HDV version of the video in the future, and in the meantime easily down convert to DV, to edit and produce a standard definition DV version right now without affecting the HDV content on the tape.

Aside from the Firewire output option, the camera also enables you to view any recorded video via composite video, S-Video, and component video.

Another significant improvement compared with the earlier JVC cameras is the inclusion of three, 1/3" one mega pixel CCD chips (image sensors). Devoting one CCD to each of the primary colors (red, green and blue) produces richer, vibrant and more accurate colors. The only slight compromise that Sony made in this setup is that each CCD has a resolution of 960x1080.

As you might have noticed, the horizontal figure of 960 does not match the full resolution of the HDV spec (1440). It seems as though this was a compromise that Sony had to make in order to work at this higher resolution of 1080 horizontal lines. However, in a real world setting, it's hard to notice this difference because the camcorder produces such stunning quality footage.

Image courtesy of Sony Electronics Inc

Yet another winning move for Sony was the choice to use a Carl Zeiss Vario-Sonnar® 12x optical zoom lens, with an anti-reflective coating and optical image stabilization. Although the lens is not removable, Zeiss is a name that is synonymous with quality, and as already stated, the image it produces is simply superb. The lens diameter is significantly larger than any previous DV camcorder, so if you decide to purchase this camera you will probably need to invest in new 72mm diameter filters (unless you own a suitable matte box). Also, your first purchase should be a very high quality skylight or similar filter to protect the lens from accidental scratches, etc.

The focal length of the lens has a range of 4.5mm to 54mm and the aperture ranges from f/1.6 to f/2.8. Incidentally, a very nice touch is the inclusion of a metal iris knob near the front of the body— this is far more ergonomic than a push button. Shutter speed ranges from 1/10,000 of a second to ¼ of a second, with some constraints depending on the shooting mode selected.

For users such as wedding and event videographers, the low light capability of any camcorder is of critical importance. The FX1 does not disappoint, with its specifications reading 3 lux as the minimum illumination. Real world tests indicate that it is not quite as good as the Sony VX2000 or PD150 in this regard, but it's not far off.

Zoom controls are selectable as manual or automatic; the manual mode uses an ergonomic ring on the lens itself, enabling more accurate handling and the option of a very fast or very slow zoom motion.

Focus is selectable as manual or automatic. The automatic mode is one of the most accurate you will come across in any camera, with little of the 'hunting' that so often makes auto-focus modes in other cameras unusable. The manual focus ring on the lens is easy to use, and is conveniently marked with distance indicators as a

Sony HDR-FX1

reference guide to allow consistent focusing on shots that need to be repeated several times, such as when pulling focus.

The 16:9 aspect color LCD monitor has a resolution of 250,000 pixels and is mounted on top of the camera. Not only is the widescreen format a wise choice, the bold and innovative decision to relocate the monitor from its traditional place on the side to the top works extremely well. Once the monitor has been flipped open the VCR transport controls come into view, along with buttons to control a few other functions.

An interesting and useful feature of the camera is the ability in HDV mode to press a button and zoom into the image for several seconds, without losing any quality. Since an HD image contains far more pixels than the LCD monitor is able to display, this zoom mode enables more accurate focusing when recording in HDV mode. When the button is pressed, the image on screen is magnified by a factor of 4x, so it appears as though you have zoomed in on the image. You can then tweak the focus and be confident that you have the sharpest possible image. The button does not work while in DV mode or while recording, only when the camcorder is in record-pause mode.

In addition to the decent sized flip-out LCD monitor, the usual viewfinder monitor is included; in this case it is a color unit with the option of an attachable eyepiece or eyecup, both of which are interchangeable according to your preference. In the area where the LCD monitor would usually sit on a Mini-DV camera, that area now houses the tape compartment door.

Although it has been criticised by some filmmakers for not having the capability to shoot in native 24p (24 frames per second progressive), the FX1 does have some 'pseudo' film modes that mimic the look of film. Dubbed 'CineFrame' by Sony, the two modes are CineFrame 24 and CineFrame 30 (the European version of the camera has CineFrame 25). The camera also uses what Sony calls a CinemaTone mode in combination with the CineFrame settings.

Without getting into complex technical explanations, the modes try and adjust the picture (particularly the gamma) and frame rate settings, to achieve a look similar to film. When played back on a TV monitor these settings work surprisingly well, although they are not designed to be a miraculous substitute for actually shooting in 24 frames per second. Cameras like the JVC HD-100U do that much better. The main problem with the CineFrame modes is that they reduce the vertical sharpness of the image.

Audio

The FX1 has an on-board stereo microphone which is acceptable for recording ambience. Of course, any on-board microphone is always a compromise since it invariably picks up some handling noise of the unit and/or user. An external audio input is provided via a 1/8" (3.5mm) stereo mini jack socket. Audio levels can be controlled automatically via the AGC system (automatic gain control), or manually using a thumb-wheel at the rear of the camera. A small headphone jack is conveniently located near the side handle for monitoring.

So what else is useful about the FX1? Well the list of features really is a long one.

You know the annoying way that clip-on lens caps dangle from a little cord and flap around in the wind, even flying in front of the lens to ruin a shot? Yes you could fix it in place behind the hand strap, but that takes extra time and effort with no guarantee it will stay in place. Some clever people at Sony acknowledged this problem because the new style lens hood on the FX1 is quite ingenious. It is actually a lens hood and cap all in one. With the flick of a lever on the side of the hood, protective horizontal flaps swing open or close, to reveal or hide the lens. It's a simple idea, perfectly implemented, and it works well. No more fumbling around with the lens cap...

Equally as useful are the two built-in neutral density filters. They really are built-in, inside the camera. If you are unfamiliar with neutral density filters, they reduce the amount of light coming in through the lens without affecting any other aspect of the image. For example, if you were shooting on a very bright day, you might need to reduce the amount of light entering the lens for the optimum exposure, and that's where a neutral density filter comes into play.

The two built-in neutral density filters have several important advantages compared with screw-on filters. First of all, they are protected from dust, fingerprints, scratches and other damage. Secondly, they can be implemented virtually instantly, at the touch of a button. The camcorder actually constantly senses the amount of light coming into the lens and it indicates on the viewfinder when the ND1 or ND2 filter should be used. Why two filters? Each one offers varying degrees of light reduction, ND2 being the greatest reduction of the two.

Image courtesy of Sony Electronics Inc

How to Shoot, Edit and Distribute HDV

Three 'assign' buttons allow you to link a certain function you use frequently to a button, kind of like a macro that you might setup on a computer—there are three buttons so you can assign your three favorite functions. It might be backlight, spotlight, white balance etc. The assigning is done through the menu.

Another useful inclusion is an extra zoom rocker on the top of the camcorder, as per the VX2100 and PD170 camcorders. This is very useful for situations where you are operating the camera in such a way that you are looking down on it, or are holding it at low level near the ground. Probably due to the fact that the rocker is fairly small and hence a little fiddly to operate, a separate switch allows you to select the zoom speed—low, high, or variable.

Batteries

Last but not least, Sony to their credit has done something which will make a lot of VX1000 / VX2000 / VX2100 / PD150 and PD170 camera owners very happy. They have used the same sizes and specs of batteries for the FX1, so your batteries from these older DV cameras can be used with the FX1. Now that's a very smart thing to do.

Incidentally, if you are tempted to buy cheap, imitation batteries for the FX1, don't. They won't work. Genuine Sony batteries use a clever technology called Info-Lithium, which gives you a read-out in the LCD display as to how much usable time is remaining on the battery. Most cheap imitation batteries do not include this technology and the FX1 will simply refuse to recognize these batteries.

Although the Sony FX1 does not have a still image capture facility, it is relatively straightforward to capture and export a single frame from recorded video using your favorite video editing software, so the lack of an onboard media card for that purpose is a minor inconvenience.

One downside that some users have reported with this camera is that some matte boxes do not fit on the front of the lens because the built-in microphone is positioned too far forward. However, matte box manufacturers were quick to recognize this and suitable, redesigned matte boxes can now be purchased from a variety of sources.

Accessories

A range of official Sony accessories are available, such as the hard carrying case (Model LCH-FX1), a wide angle lens (Model VCL-HG0872), a high capacity battery (Model NP-F970, and a retractable shoulder brace (Model VCT-FXA). However, a lot of people have commented about the outrageously high prices for some of these accessories, so its worth shopping around before you buy.

On its release, this camera had a retail price of around $3700, and at the time of writing it was available for a street price of approximately $3000.

Chapter 8
Sony HVR-Z1 (U suffix for USA region, E for Europe)

If the FX1 is the camcorder that put HDV on the map, then the Z1 is the camcorder that secured its place there and put Sony out at the forefront of HDV capturing devices. The Z1 Camcorder is the professional version of the FX1 camcorder, in a similar fashion to the Sony VX2100/PD170 relationship, so as you might expect there are a lot of similarities, but there are also some major differences.

Many professionals are comparing this camera favourably to DigiBeta, a startling comparison considering the price difference, but it shows the confidence that the camera has instilled.

Sony Z1U shown with a Sony wireless microphone receiver
Image courtesy of Sony Electronics Inc

Sony HVR-Z1

Interestingly, when the camera was first announced, numerous individuals posted comments on the various online forums that demonstrated an amazing lack of insight. The widely held view was that the only major difference with the Z1 compared with the FX1 was that it had dual XLR inputs for audio rather than the mini jack on the FX1. These same people went on to say that they would not be parting with the approximately $1500 difference in price compared with the FX1, just to obtain XLR inputs. How misinformed they were!

Of course that's not to understate the importance of the XLR inputs, they are crucial for a professional camera. In addition, Sony has provided two channels of inputs, each with the facility for phantom power, and each with independent gain controls on well designed thumbwheels at the rear of the camera. These are neatly protected by a removable cover to prevent the levels being accidentally altered once they are set. Another thoughtful detail.

Note the XLR connectors on the side of the lens barrel
Image courtesy of Sony Electronics Inc

While the XLR inputs are the most obvious cosmetic difference when looking at the external appearance of the Z1, there are many other valuable differences that are unseen. In fact, there are some *forty* differences between the FX1 and Z1 models, so the price difference is definitely justified.

34

How to Shoot, Edit and Distribute HDV

What seems most surprising is that in all the countless reviews of the Z1 that have appeared since its release, only a handful of reviewers pointed out the significance of the Z1's capability to shoot both PAL and NTSC DV formats, and HDV at 50i and 60i! This ability is truly unique, a first, and a major milestone in the world of video cameras. Anyone who travels to different countries to shoot video locally will find the Z1 to be an answer to a prayer. You will also benefit from this feature if you are located in the USA and intend to output your project to film. Why? You can now record at 25fps at the higher line resolution that PAL affords.

A Shooting Tip

If you are shooting in a fixed location and you need the highest possible recording quality from the FX1 or Z1 cameras, connect the camera's component output to a portable HDCAM recorder. Since the heavy MPEG-2 compression of the HDV format is only applied when video is recorded to the Mini-DV tape, you can benefit from the highest quality, component output prior to that stage. Obviously it's not as good as a true HD-SDI output, but the component signal will do the job nicely for all but the most demanding of applications.

Again, the Z1 allows you to shoot and playback DV and HDV in either 25fps or 30fps (50i or 60i), and in PAL or NTSC DV, all on one camera, simply by selecting the appropriate setting on the menu and then switching the camera off and on again (the whole process takes seconds). It really is like having multiple cameras in one.

The 'CineFrame' film-look modes of 24, 25, and 30fps are also included for even more versatility. As with the FX1, these modes aim to imitate the look of film. Should you use these modes? It depends. Bearing in mind that the camera discards some data when using these modes, if your intention really is to output to film then you should not use the CineFrame mode. Specialist external processing at the editing stage will yield better results and you'll have more data available to feed into the process. However, if you simply want to create a film look to your video then these modes produce acceptable results.

Ultimately whether you use the CineFrame 24, 25, or 30 mode is down to your personal preference. Why not experiment to see what works best for you?

The 'Elusive' Film Look

If you are really keen on the film look, you may want to try shooting in 25fps, rather than 30fps. The Sony Z1 allows you to select 25fps (50i). Don't forget you'll need to use a PAL monitor during your shoot! The thinking behind using 25fps is that it's so close to 24fps that you can use a software converter such as Apple's Cinema Tools Suite that is integrated with Final Cut Pro. This software allows you to convert your 25fps/50i footage to 24 progressive. Essentially what you are doing is removing one frame per-second and slowing down the audio by about 4%. Remember, DVD's can playback 24p video, and it looks great on screens that are optimised for progressive video signals, which most modern TV monitors are. Another advantage of using 24fps on your DVD is that you'll be losing around 20% of the frames compared with 30fps, so your programme will be a smaller file size. That means you can use less aggressive compression for even better image quality. You can read more about the 'film look' later in this book.

What are some of the other differences of the Z1 compared with the FX1 model? In addition to the decent sized flip-out LCD monitor, a viewfinder monitor is included; in this case (unlike the FX1) it is selectable as color or black and white. The viewfinder

How to Shoot, Edit and Distribute HDV

can be on at the same time as the LCD screen, another feature not found on the FX1.

The Z1 also has the option to record both DV (short play only) and HDV in DVCAM mode. DVCAM mode is preferable to DV mode because the tape runs faster, meaning there is more tape to record the same amount of information, which in turn reduces the likelihood of errors or dropout. The downside is that you can't fit as much footage onto the tape. For example, a 60 minute Mini-DV tape records about 43 minutes of footage in DVCAM mode.

The component output is more versatile with the Z1, allowing any format that can be recorded to be output as a component signal, with additional options to squeeze, letterbox or edge crop the image according to your needs and the type of monitor you are using. This is a very versatile and powerful feature.

Where the FX1 has three 'assign' buttons that can be user programmed to turn your favorite settings on and off literally at the touch of a button (white balance, backlight etc), the Z1 has a total of six of these buttons for greater flexibility and choice.

The 'Picture Profile' menu also allows you to go wild with customization. This is where you will gain access to the CineFrame modes plus a lot more. Although the FX1 has this menu too, the options on the Z1 are more comprehensive. You can make changes to color level, color phase, sharpness, skin tone detail, skin tone level, auto exposure shift, AGC limit (automatic gain control), AT iris limit, white balance shift, ATW sensitivity, black stretch, and the CinemaTone modes previously mentioned.

Obviously these options allow for many creative shooting possibilities, and you can store, copy and paste combinations of settings as personal presets within the picture profile menu. The camera is very highly recommended as an ideal all rounder.

Chapter 9
JVC GY-HD100

JVC launched the HD100 camcorder at NAB 2005 (the National Association of Broadcasters convention), and it was available for purchase from late 2005.

Although it falls within the professional rather than consumer bracket, something that left many people scratching their heads was JVC's decision to refer to this camera as 'ProHD', even though it records HDV and conforms to the general specification and definition of HDV. ProHD sounds like a new, different format, but it's NOT. Whatever the reason, the move seems to have caused confusion in the marketplace.

Leaving the controversy of the name or branding of the camcorder aside, is it any good? Since it was introduced after the Sony Z1, does it have any additional or different features? In a word, yes. JVC has done a great job at offering some unique differences to enable it to compete with the Sony Z1.

Image courtesy of JVC

How to Shoot, Edit and Distribute HDV

The two major differences are the lens and the ability to record 24fps. Let's take the lens first. The camera has the ability to change lenses according to the shooting situation unlike the fixed lens system on the Sony HDV cameras.

Although the HD100 ships with a 16 x Fujinon lens and the standard lens mount is a 1/3" bayonet, other compatible lenses can be used, including ½" lenses. Many professional film makers insist on the ability to change lenses, so this will be a very attractive feature for them.

Let's talk further about the included lens. It's quite remarkable that JVC included the Fujinon lens with this camera. Many people are wondering how JVC could include an HD lens with the camera for a total price of around $6000, when the lens on its own in theory should cost far more than that amount. Perhaps volume production enabled them to do it?

The other major difference is the ability to record 24fps progressive. Many film makers had been eagerly awaiting this feature ever since the HDV format was first introduced. Earlier we mentioned why 24fps is considered such a big deal among certain users.

Again to remind you, the 'film look' is something that is difficult to describe, but many directors and videographers have long sought to make video look like film. 24fps is the frame rate that is used to shoot motion pictures, so with the JVC HD100U, you can now shoot High Definition video in 24fps and subsequently transfer the footage to 16 or 35mm film more easily and cheaply than with other video frame rates.

24fps is also considered an attractive recording mode because even when it is played back on a TV or monitor, it more closely resembles the look of film than the 30fps used by NTSC video or

39

even the 25fps commonly used in the PAL video format. You would think that a higher frame rate would result in better quality, but for some reason it seems to be the general consensus of opinion that 24fps is the magic number that looks the best, even though it slightly stutters compared with faster frame rates. Clearly a large part of the perception is down to the psychology of how our brains interpret different frame rates.

Why Use Interchangeable Lenses?

Why is the ability to change lenses so important to some camera operators? A fixed lens is a compromise designed for general shooting environments, and it simply can't do it all. For example, wildlife photography usually requires long zoom capabilities, as does shooting sports events. Likewise, if you were shooting nature photography, you would probably require a high quality macro lens for extreme close-ups. As you can see, the ability to change lenses opens up many more shooting possibilities.

The 'film look' is a complex subject that could easily fill a separate book, but suffice it to say there are a number of factors that contribute to reproducing the 'film look' with video—if you are interested in this you should know that there is much more to it than simply shooting in 24fps. Other factors such as lighting, shot composition, makeup, and post production techniques all play an important part in making video look like film. We'll cover some of these later in the book.

Something else that JVC have highlighted in their promotion of the HD100U camcorder is that the MPEG-2 compression is less 'aggressive' than that used by the Sony HDV camcorders. The main reason for this is that the JVC shoots 1280x720, which is a lower resolution than the 1440x1080 that the Sony cameras use. Of course, 720 lines is still High Definition. Although the HD100 is technically lower in resolution than the Sony, the lower compression ratio (a GoP of six frames) that the JVC unit records is something of an advantage when it comes to compression.

Another feature of the JVC camera is four channel audio recording capabilities, with independent sectors for MPEG-1 layer II tracks, and 16-bit, 48kHz PCM Audio which is uncompressed in a dedicated area of the track.

The HD-100 was designed as a professional, shoulder mounted camera.
Image courtesy of JVC

An SD memory storage card at the rear of the unit allows the user to store various camera settings. This is a handy feature that could be used to transfer settings to another HD100 camera, or enable you to be instantly ready with a specific setup or 'look' when shooting in the field.

JVC GY-HD100

An earpiece is ergonomically placed next to the viewfinder to allow for easy monitoring of the audio.

The built-in 230,000 pixel viewfinder can be moved in a range of directions so that it sits comfortably for the user. The 250,000 pixel 3.5" color LCD screen is in the traditional flip-out area on the side of the camera, and the on-screen displays and menu functions can be turned on or off as with most other cameras.

An interesting function called 'focus assist' is provided to help overcome the challenge of focusing when shooting in HD mode. It differs greatly from Sony's approach. When the focus assist mode is enabled, the viewfinder turns the image to black and white and the viewfinder adds a vivid color around the edges of the subject or object on screen when it is in focus. It makes it easy for the user to see at a glance whether a particular subject is in focus.

All in all, JVC to their credit have managed to effectively appeal to a different corner of the HDV market with this camera. It's a camera for the serious professional that needs to change lenses, intends to have their work easily transferred to 16mm or 35mm film, or who desires to imitate the look of film with video.

On release, the camera had a recommended retail price of around $ 6295, but it can be purchased from many dealers at a lower 'street' price.

The HD-100 shown in use when mounted on a tripod.
Image courtesy of JVC

Chapter 10
Sony HVR-A1

No doubt encouraged by the phenomenal success of the Z1, Sony followed up quickly by releasing a smaller sibling, the HVR-A1. Designed for professional users, this surprisingly compact camera, weighing just 1.5lbs, uses a single chip CMOS design, a radical departure from the CCD technology we've become used to in recent years.

Today's CMOS technology is a world apart from its widespread use in the analog camcorders of years ago. In fact, the image quality on the A1 is excellent, especially considering its compact size. No doubt some of the credit should go to Sony's enhanced image processor (EIP) which works in partnership with the

Image courtesy of Sony Electronics

CMOS to produce the best quality image. Admittedly, using a CMOS sensor is not as good as using 3 CCD's, so don't expect miracles, and you may experience unusual images from time to

Sony HVR-A1

time due to the rolling shutter technology. Unlike CCD's, which capture full frames at once, the CMOS technology in this camera only captures consecutive lines, which leads to some odd looking video when performing a fast pan or tilt. It's a current limitation of this technology, but it should not present too much of a problem for you if you shoot according to that knowledge.

The Sony A1 and HC1 use CMOS rather than CCD image sensors.
Image courtesy of Sony Electronics Inc

Optics obviously play an important role and as is their custom of late, Sony contracted Carl Zeiss to come up with a 5.1mm to 55mm vario-sonnar zoom lens (10x). Interestingly, Sony says that due to the unusually high dynamic range of the CMOS sensor, the A1 did not need to include any neutral density filters like the Z1 and FX1. As a result, in bright lighting conditions, images recorded on the A1 look exceptional.

Despite its small size, many features on the much larger Z1 are also present on the A1, including DV and DVCAM recording modes, down-conversion to SD, timecode options, and versatile viewfinder options.

The LCD screen is placed in the conventional position on the side of the camera, and is 2.7" wide, with 123,000 pixels in the 16:9 format. Similar to the FX1 and Z1's CineFrame mode, the A1 has a 'cinema' mode that aims to give a 24p look to the video. The results are a matter of taste. Slightly stuttery video is not for everyone.

Professional audio is not neglected with the option of an add-on module that sits on top of the camera to provide two XLR inputs with phantom power capability. Something that the A1 features that is not present on the Z1 is what Sony calls 'Super Night Shot.' As the name implies, it allows you to get super shots at night!

How to Shoot, Edit and Distribute HDV

Realistically, there is never any substitute for good lighting, but if the circumstances give you no option, this mode can at least allow you to capture some type of image, albeit with a green cast.

There are a couple of downsides to the camera that are worth noting. One is the decision to utilise a bottom loading tape mechanism. Whoever dreamed this idea up should be severely reprimanded! It becomes a real issue when using the camera on a tripod, as it is simply not practical to change tapes—you can't open the cassette compartment!

The camera uses yet another type of Sony battery, you cannot use the Z1 or FX1 batteries with this unit—another irritation; but given the small size of the unit it's understandable because those batteries would probably be too bulky. Incidentally, the standard supplied battery is hopelessly inadequate in its running time; no doubt Sony wants to encourage users to purchase a larger capacity battery, which you will almost certainly have to do.

Lastly, the menu functions have to be accessed by touching the flip out LCD screen. Not only is this a little fiddly, but constantly touching the screen with greasy fingers does not sound like a good idea for any serious shooter.

Where does this camcorder fit in the real world of professional video? Anyone who needs to shoot HDV inconspicuously while retaining the ability to make manual adjustments would certainly be able to do that with this camera. For example, there are many shooting locations that would prevent you from entering with a professional looking camera, whereas if you were carrying this camera (especially without the audio module on top) it is so small that it easily passes as an inexpensive consumer model.

Additionally, many wedding and event videographers who are looking for a second or third HDV camera could probably afford to purchase this camera rather than another Z1 or FX1.

Chapter 11
Sony HDR-HC1

In promoting the HDR-HC1 as the first HDV camcorder under $2,000, Sony achieved another milestone in the short history of HDV. Any camcorder under $2000 will open itself up to the mass market, and that's exactly what Sony intended with the HC1. The timing was perfect-at a point when consumers were becoming increasingly aware of the benefits of High Definition. So with the HC1, HDV for the masses had truly arrived.

As a general guide, the HC1 is the same basic camcorder as the A1, with some differences to gear it more for the consumer market. On that point, semi-pros and even some pros should not dismiss this camcorder. Given its fairly wide range of manual controls it may be suitable for you in certain situations.

Overview

The camera uses a single 4:3 aspect ratio CMOS chip, although the way the camera works with the image means it can record native 16:9. As with Sony's A1 camcorder, the CMOS image sensing system is used. In good lighting conditions the HC1 produces excellent quality images, with rich vibrant colors and stunning High Definition when played back on a compatible HD monitor or TV.

Design / Layout

One thing that strikes you when looking at the camera is that the lens dominates the body of the unit. It's been mentioned before, but HDV demands great optics so it's understandable that the lens is large compared with DV camcorders.

Ergonomically, the camera is as well designed as any modern Sony camera, and clearly a lot of thought has gone into placing the various buttons, dials and displays in the most logical places for the user, rather than simply doing what's best from a design or visual point of view.

Image courtesy of Sony Electronics Inc

Something that aids the uncluttered approach is the touch screen menu system on the flip-out LCD screen. The same comment made for the A1 equally applies with this camcorder, namely that the down side to this approach is that your greasy fingers are constantly touching the LCD screen.

The camera is small enough to be held in one hand, and is comfortable and nicely balanced. In that sense it's a delight to work with.

Still Capabilities

Given that the HC1 is aimed at the consumer market, it's no surprise to see that the camera has good still image capabilities. Still images can be captured in either 4:3 or 16:9 aspect ratios, up to a maximum resolution of 1920 x 1440 pixels.

Sony HDR-HC1

There's a pop-up flash above the lens which can be set to low, medium or high, and the camera records stills at a resolution of 2.8 megapixels on to a Memory Stick Duo, which is half the size of a standard Memory Stick. Also, the HC1 has a burst mode, which allows you to take up to 25 images consecutively, which may be useful for capturing action stills or fast moving scenes.

Thankfully, Sony implemented their unique 'expanded focus' concept on this camera, which makes focusing so much easier. When you press the button, the LCD and viewfinder zoom in further (closer) so that you can view a small section of the entire image at full resolution. This allows for easier focusing.

Image courtesy of Sony Electronics

Audio

Audio is surprisingly well supported on this camera. Although the microphone input is a mini-jack, audio levels for both the externally connected and built-in microphone can be manually adjusted. That's unheard of in consumer cameras, and it's a smart move by Sony. It makes it far more usable for semi-pros and pros.

One weakness on the camera is the small zoom rocker. If your hands are large you will struggle to accurately control the speed of the zoom, especially if you want to zoom very slowly.

Another irritation, as with the A1, is the bottom loading cassette compartment for the reasons already noted. Unbelievable.

Overall, apart from the gripes mentioned, Sony has another winner on their hands. As the most inexpensive option currently available for capturing HDV, it will sell like hot cakes.

Chapter 12
Canon XL-H1

As the most recent manufacturer to launch an HDV camcorder, Canon had ample opportunity to study both the Sony and JVC HDV camcorders, and with the release of their XL-H1 HDV camera, they wisely offered some unique differences to make it stand out in the marketplace.

All camcorder manufacturers develop a significant band of loyal users, and as a result they are usually reluctant to depart from a previously successful camera design. This is apparent with the XL-H1, as it looks remarkably similar to the popular XL2 DV camera. Users of the XL2 will be reassured to know that, as they can quickly become familiar with the similar layout and placement of controls on the XL-H1. In fact, if anything the XL-H1 offers some improvements over the XL2 in terms of ergonomics and balance, even though it is slightly heavier.

Image courtesy of Canon

Launched in November 2005, the camera initially sold for around $9,000, making it the most expensive HDV camera to date.

Three CCD's has become the norm on professional HDV camcorders, and the XL-H1 is no exception. It uses three 1/3", 1.67 mega pixel interlaced CCD's (native 16:9) at 1440x1080 resolution. Although the 1080/60i recording method is used, Canon uses an open architecture system that allows you to select a frame rate of 30 or 24.

The 24'f' setting is purely a marketing term, and is not true 24 frame progressive recording, although visually it seems to be more widely praised than Sony's CineFrame options. Canon says that once captured into your editing software, you won't be able to tell the difference between their 24f system and cameras that natively capture 24p. Tests on that claim seem to be subjective at best and inconclusive—some users agree, while others dispute the claim. Canon are sticking to their guns and they have tried to emphasise the distinction by referring to their 'movie mode' as 24f (f for frame). They can't use the term 24p because it's not true native 24 frame progressive recording.

HD-SDI Output

The single most important feature of this camera that gives it a unique selling point in the marketplace is its ability to output uncompressed 1080 (60i) by means of an HD-SDI connection. This feature alone will attract many pro users to this camera. SD-SDI output is also possible (for standard definition).

What is HD-SDI?

HD-SDI is a very high quality digital signal format used in high-end video gear. For example with the Canon XL-H1 it allows you to connect the camera to an HDCAM or DVCPRO HD deck to record *uncompressed* High Definition images directly from the camera, rather than recording to Mini-DV tape in the compressed HDV format.

In real terms, the HD-SDI capability means that the XL-H1 is more accurately described as an HD camera *and* HDV camcorder.

How to Shoot, Edit and Distribute HDV

Can you see why that is? Since the XL-H1 can output uncompressed HD-SDI, it could be used in an HD broadcast environment, either as an HD studio camera, or out in the field with a direct connection to an outside broadcast truck. At the same time, it also provides the option to record HDV onto a Mini-DV tape.

An interesting add-on option is Canon's 'Console Image Control Software'. It controls camera settings remotely, and provides image analysis tools in a similar fashion to Serious Magic's DV Rack. If you've ever used a CCU (camera control unit) then in essence the Console Image Control software provides similar functionality. The main difference is the software interface (rather than hardware) and the fact that the control signals are fed through the Firewire cable rather than a separate control cable.

Optics

The XL-H1 lens is interchangeable with standard Canon lenses, a benefit that many pro users cite as essential. The supplied lightweight 20x HD lens is excellent, (5.4mm–108mm), with an aperture ranging from f/1.6–f/3.5. Note that the 20x zoom capability is greater than other HDV camcorders.

The excellent optical image stabilizing system used on this and many other Canon products is arguably the best in the business. Without getting bogged down with technical details, the advantage with the Canon system is that the image is examined further at the CCD stage to detect and correct low-frequency vibrations that were missed by the gyro. The two step process allows for extraordinarily accurate image stabilization.

Fine image control adjustments are possible, which will be particularly useful to anyone who wants to get a consistent look to their production throughout multiple shooting locations. There are three color matrixes for a wide range of color correction options, and two cine gammas for intricate adjustment of dynamic range. Likewise, knee, black stretch, horizontal detail, coring, sharpness, color gain, hue, and overall color can be adjusted, all independently.

Still images can be captured using the 'photo' button at the full resolution of 1920x1080, to an SD or MMC memory card. This is equivalent to about 2 megapixels. A burst rate of up to five frames per second adds further functionality to this feature. The memory card can also be used to store camera settings, with the facility to transfer those settings to another XL-H1 if desired.

Image courtesy of Canon

If you plan to use multiple XL-H1 cameras in one location, the ability to transfer camera settings would save considerable setup time—you can simply insert the memory card into each camera and copy over the required settings. Up to six combinations of settings can be stored in the camera for quick recall.

The viewfinder and LCD monitor are combined in one unit as on earlier DV models, using a 2.4" color screen, in a 16:9 aspect ratio, with safe area markings built-in. It can also be used in black & white mode.

To use the viewfinder as an LCD monitor, you simply press a release button under the viewfinder which flips the magnifying part of the viewfinder open to reveal the LCD screen.

Aspect ratio guides can be turned on in the viewfinder, helpfully displaying a comprehensive choice of 4x3, 13x9, 14x9, 1.66:1, 1.75:1, 1.85:1, & 2.35:1 composition guides within the 16:9 frame.

To aid in focusing, Canon have adopted a combination and variation of both the Sony and JVC approaches. Within the viewfinder, one option (misleadingly called 'peaking') displays an exaggerated line until the image is focused. The second option, a button named 'magnify' zooms into the image temporarily, much like the Sony 'expanded focus' button.

If you have ever tried to shoot computer monitors, perhaps in an office environment, you'll know that the different scan rate of monitors shows up on your video camera as horizontal lines continuously scrolling up the computer screens. This distracting annoyance can be overcome with Canon's 'clear scan' feature. It lets you match the scan rate of the monitor with the scan rate of the camera. It's a neat feature that more manufacturers should offer.

Inputs and outputs far exceed what you might expect on a camera of this price. For audio there are two XLR inputs with independent audio control of each channel, a microphone jack and a headphone jack. Then there are BNC jacks for gunlock-in and timecode in and out, the HDI/SDI already mentioned, as well as composite video, component video, and S-Video out. The Firewire port supports HDV and DV.

This is clearly a camera for the experienced professional who likes to shoot on the shoulder, or who needs the HD-SDI capabilities and ability to use interchangeable lenses. While other users would obtain good footage from it, they would probably be paying for a number of features that they would not be likely to use.

Chapter 13
Buying Advice

If you are in the market for an HDV camcorder, what model should you choose? Ultimately, it depends on what type of shooting you plan to do, the level of experience you posses, and your shooting style.

As you've hopefully seen in the preceding reviews, each manufacturer has geared their HDV products to different target markets. Weigh up the features that you feel you will most benefit from, and then see which camera fulfils those criteria.

Obviously, budget plays a part too. It's important not to lose sight of the fact that the whole point of the HDV format is to enable you to record High Definition in a cost-effective way. The problem with that is that everyone has a different idea of what is considered 'cost-effective.' Hence, if your ideal camera is out of your price range, you can still step into the world of HDV with a cheaper but capable HDV camera.

Beware of Dodgy Dealers

While we're on the subject of spending your hard-earned money, please don't get sucked in by one of the many advertisers on the Internet who offer HDV camcorders at prices that seem too good to be true. If it sounds too good to be true, it probably is.

The mark-up (profit) on most electronics is fairly small, so it's simply not possible for any legitimate company to offer prices that are considerably cheaper than the norm. Many New York establishments in particular have developed a bad name for shady business practices. They have been known to attract customers with amazing offers for Product X, only to cleverly

How to Shoot, Edit and Distribute HDV

manipulate them into buying Product Y, on which they make far more profit. Another tactic is to sell camcorders without any power supply, cables, manuals, or accessories.

A further scam is to take payment for a camcorder over the Internet or over the phone, with no intention of fulfilling the order for that product. When you contact the company to find out why your product has not been delivered, they will try and force you into exchanging your order for something else, or only fulfil the initial order if you agree to buy overpriced accessories.

The aim of telling you this is not to scare you, but to make you aware and hopefully prevent you becoming a victim. It's sad that the authorities do not do more to drive these unscrupulous dealers out of business.

A notable exception is *B&H Video* in Manhattan. They are authorised dealers for various manufacturers and they offer fair prices and good service. Likewise, Edgewise Media based in California offer both competitive prices and outstanding customer service. These companies will take care of you. Of course, they are not alone.

If you are in any doubt about a particular dealer, there is an excellent free service on the Internet which allows you to check independent reviews of a company by individuals who have used them, at: www.resellerratings.com. Read a few reviews and look at the ratings and you'll soon see if a company has a good reputation or not.

If you enjoy getting the best deal on your equipment and software purchases, check out: www.techbargains.com. It's a daily updated list of computer and electronics items that are on sale in stores or on the Internet. If you are considering a purchase, I highly recommend checking out this useful resource before you buy, but always remember to check that the seller has a good reputation.

Chapter 14
Shooting HDV

Right at the outset let's establish that shooting HDV has significant differences compared with DV. Of course, if it came down to it you could pick up an HDV camera and record some satisfactory footage using the automatic settings on the camera, but there is much more to professional shooting with HDV. And you want to be professional, don't you?!

Is it 'Shooting' or 'Filming'?

Why do we refer to 'shooting' rather than 'filming' in this book? Technically, it's correct to use the term 'shooting' when referring to video because 'filming' relates to film. Having said that, it's acknowledged that the term 'filming' is used so frequently these days that most people think of filming as any form of capturing moving pictures, whether it be with a video camera or a film camera. In fact, it's even common to hear the term 'video filming' being widely used, which is really an oxymoron.

In a similar way, in some countries people often refer to using a 'hoover' on their carpet, even when they are using a different brand of cleaner. Again, it's more accurate to use the term 'vacuum cleaner'. However, since Hoover was the predominant brand of carpet cleaner for so many years, that's the term that became rooted in the minds of many people. The same thing happened with the term 'filming' to the point where it's now used almost universally to describe capturing moving images with any device.

Still, it seems appropriate to use the technically correct term in this book, so now you know...

Widescreen Aspect Ratio

One of the most immediately apparent visual differences when viewing HDV is the widescreen (16:9) aspect ratio. If you've worked with DV for years in the 4:3 aspect ratio, it will take you some time and practice to adjust to shooting 16:9.

A traditional 4:3 aspect ratio TV compared with the popular widescreen 16:9 format

Shot composition is quite different with 16:9. For example, if you have a single interview subject on screen, where would you place them in the frame? Conventional 4:3 shot composition would place them to the far left or right of the screen with 'looking space' on the other side of the frame. By contrast, with a 16:9 shot, the subject would probably be better placed just left or right of center (still with appropriate looking space), rather than at the very edge of the frame.

Another aspect of 16:9 composition that requires getting used to is the ability to see more in the frame. For example, landscapes look very different and much more dramatic in widescreen. Likewise, a conversation between two people can be comfortably shown in one frame rather than having to cut back and forth between them.

Arguably, there is no better way to learn widescreen composition than to study feature films from well known Directors to see how

Shooting HDV

they handle different subjects. Look for tips and inspiration from popular and classic films. Try and identify what types of shot are particularly effective or dramatic in the 16:9 aspect ratio. Once you have a grasp of the basic techniques by learning from watching movies, ultimately there is no substitute for real world practice shooting, so get plenty of experience before your first paid shoot.

> **An Important Framing Tip**
>
> If your camcorder allows it, set the LCD screen or viewfinder to display the safe area guides for 4:3 on screen, while you shoot in 16:9. Why? While you are framing your shots for 16:9, you'll also be aware of the 4:3 shot composition, and if the project eventually has to be output to a 4:3 version, your composition will still be fine.

Focusing

When shooting DV, focus is fairly forgiving, you can often get away with a shot that is very slightly out of focus. Not so with HDV. Be warned! This is one of the most significant and important differences in shooting HDV compared with DV. A shot that is slightly out of focus in HDV (or HD) is glaringly obvious, and it looks awful.

If you have access to HD channels on your TV, watch virtually any HD program and you will see examples of this. It's particularly evident on live or near live TV shows such as the late night chat shows or reality shows that are filmed in HD. What does this tell you? If highly trained camera operators working for major broadcasters have a challenge focusing in HD, you can see that it's going to be equally difficult for the rest of us. The fact is, it is considerably more difficult to focus with HDV than it is with DV, and it will take you considerable practice.

How to Shoot, Edit and Distribute HDV

Sony, to their credit, recognized the focusing challenge before they produced their FX1 and Z1 cameras, and they incorporated a feature which allows the user at a push of a button to 'zoom in' to the frame and see a magnified area of the image at full resolution, which makes it much easier to see if a shot is in focus. Canon and JVC likewise have implemented systems on their cameras that aid focusing.

What else can you do to aid focusing? Although expensive, an external HD monitor is essential for serious shooting. You can't trust the standard display of the camcorder's small LCD flip-out screens for accurate focusing. Companies such as Marshall (www.lcdracks.com) produce some of the most affordable LCD monitors. Even a larger, external SD monitor is better than the small internal monitor on the camera, but there is no substitute for an HD monitor.

A Key Tip to Help Focus

Monitoring video in black and white makes it easier to see when your focus is sharp. All studio TV cameras monitor in black and white for this reason. Some of the more recent HDV cameras have viewfinders that are selectable as color or black and white.

If you are on a tight budget, one option you might consider for accurate monitoring in the field is a high resolution LCD monitor that accepts component inputs, such as the Dell 2405FPW (www.dell.com). Apple also make excellent 20", 23", and 30" widescreen displays in the form of their Cinema range (www.apple.com/displays), but you will need a special converter to change the component output from the camera to DVI, as the Cinema displays do not accept component video directly. StarTech's VID2DVIDTV (www.startech.com), should work for that purpose.

Shooting HDV

Panning

Perhaps the most frustrating thing about the HDV format is its weakness in displaying fast panning camera moves. MPEG-2 compression simply can't keep up with the sudden changes in the image that a fast pan produces, so they should be avoided. Otherwise the picture will simply appear to break-up with numerous digital artefacts—not a pretty sight. Always, always, *always* (yes, that is three times for emphasis) make sure your pans are slow and smooth.

Camera Moves

If you've watched a reasonable amount of HD programming on TV, you will probably have noticed that there are far less rapid camera moves with HD shows compared with standard definition programming. Very rarely will you see anything approaching MTV style camera moves. There are two main reasons why this is the case.

Firstly, although high quality HD cameras are used, 24p is often the preferred frame rate, and 24p introduces an undesirable motion jitter.

Secondly, since MPEG-2 compression is often the method of distribution for HD programming (Satellite HD as well as many cable companies use MPEG-2 to cram more channels into their bandwidth), the weakness of that format becomes apparent as the image breaks up during fast camera moves, even if it has been filmed in a high quality format such as HDCAM.

What does this mean to you? Take care to adapt your shooting style to avoid any fast camera moves wherever possible. Experiment with your camera to see what kinds of movements look acceptable.

Use a Tripod or Steadicam Device

A good quality tripod is an essential accessory when shooting HDV

This tip follows on from the previous for similar reasons. Handheld footage rarely looks great with HDV (or HD). It's difficult to pinpoint or explain the psychology of exactly why this is the case, but for now, take the advice and use a tripod or Steadicam device for all shots, as the footage will look much better. Companies such as Vinten and Manfrotto make good quality tripods within a reasonable price range.

Steadicam are famous for their 'floating camera' capabilities, and their range goes from the exceptionally lightweight Merlin device (suitable for cameras 5lb and under), up to full scale, body mounted harness systems such as the 'Mini', 'Flyer', 'SK2,' 'Clipper 2', and 'Archer'.

Another company that is prominent in the same market is Glidecam, who offer similar products at lower prices.

Pay Attention to Exposure

Any video that is overexposed looks awful, but with HDV the problem is emphasised. Keep a careful eye on your external monitor as well as on the viewfinder's zebra display at all times when shooting, to avoid clipping. If anything, it's better for an image to be slightly under exposed.

Using Filters and Lenses

If you're a creative type of shooter you probably enjoy using filters and wide angle or fish-eye lenses. When shooting HDV, don't use cheap filters or lenses because the quality of your footage will suffer dramatically. HDV cameras use very high quality optics because they need to! High quality lenses and filters are not cheap, but they are your only option if you want to retain high quality imagery. Apart from the respective manufacturers own offerings, check out products from companies such as Century Optics, 16x9, and OpTex. Century, for example, produce wide-angle adaptors, fish eye adaptors, tele-converters, and dioptres for the Sony FX1 and Z1 HDV cameras.

Direct-to-Disk Recording

Direct-to-disk recording is a major advancement in shooting technology that is available for both DV and HDV cameras. Actually, it represents a significant breakthrough in the production workflow.

What is direct-to-disk recording? As the name suggests, it's the ability to record video directly to a hard drive while you are shooting, by means of the Firewire cable. The reason it's such a revolutionary advancement is that it is an enormous time-saver.

In the traditional capture-to-tape arrangement, if you recorded a three day conference for example, you might end up with 18 hours of footage. Those eighteen tapes would then need to be captured to your editing computer. When you factor in the time spent changing tapes etc, it would take much longer than 18 hours.

Using direct-to-disk technology, as soon as you have recorded the footage you can connect the disk to your computer and you are ready to edit! What a difference! You've saved yourself three days of capturing!

Currently there are two methods that allow you to use this technology. Firstly, there are products such as the *Firestore* FS-4 range, and *CitiDisk HDV*. These products are built around a small hard drive housed in a robust enclosure. Of the two, Firestore is the more expensive option (but more widely used), whereas the Citidisk HDV is slightly smaller and cheaper.

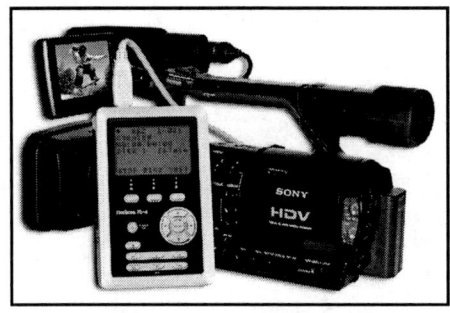

Focus Enancement's Firestore FS-4 shown against a Sony Z1 for size comparison. Image courtesy of Focus Enhancements

The Firewire cable is attached from the camera to the unit, and the video is then captured directly onto the drive, up to the drive's maximum storage capacity. The unit is usually small enough to be mounted on the camera's hot shoe, with buttons to start and stop recording, as well as to perform other basic functions such as finding how much recording time is left, etc.

Another product that comes under this first category is the Wafian HR-1, which records direct-to-disk using CineForm's compression technology. It's a larger unit, which looks somewhat like a high-end piece of hi-fi equipment. It records up to 9 hours of full resolution 1920x1080 or 1280x720 HD-SDI onto a RAID for safety. This unit is probably overkill for most HDV users but it's worth knowing about it in case a future HD project warrants its use.

The second way to benefit from direct-to-disc technology is a product called HDV Rack by Serious Magic. It's a software application that is installed on a laptop computer (or any computer for that matter), that allows you to capture video directly to the hard disk.

There are pros and cons with each of the two methods. A drive that can be mounted on the camera has the advantage of being very portable. Although you would probably find it too heavy to do sustained handheld work, you are not restricted in your movements as you would be with the Software based HDV Rack.

On the other hand, the advantage with HDV Rack is that it provides far more functionality to help you setup and acquire the best quality footage. For example, it includes an on-screen monitor

DV Rack software from Serious Magic with the HDV plug-in. Image courtesy of Serious Magic.

that can be fine tuned as a very accurate display, a color bar generator, vectorscope and waveform monitors, and live audio monitoring.

Another useful feature is the ability to constantly record up to 15 seconds into a buffer. This means that you will never miss a shot be pressing the record button too late. For example, imagine you are setup to shoot a conference but the program starts while you were distracted for a moment. Since HDVRack was continuously recording a buffer of 15 seconds anyway, when you press record it will automatically add those extra 15 seconds to the start of the recording, so you won't have missed the start after all! It's a very clever piece of software that is highly recommended.

With all of these options you can still record to tape at the same time as recording to disk. In fact, this should be done as a matter of course, as a backup measure. Computers and hard disks can be temperamental, and if you only record to disk you are risking losing your footage if it was to become corrupted somehow. We all know that even the best computers are not without their problems.

Recording Audio

As mentioned earlier, HDV uses MPEG1, layer II compression for audio. Generally, this works very well, but you need to keep a close eye (and ear) on your recording levels. Digital signals don't have any headroom once you go over 0dB, they WILL distort, unlike analog audio where there is room for peaks over 0dB.

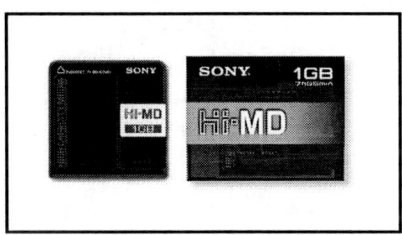
Sony's Hi-MD format for recording audio is inexpensive and versatile.
Image courtesy of Sony Corporation Inc

Likewise, you need to be careful not to let your levels record too low, because when you amplify the signal in the edit, you will be amplifying a considerable amount of noise with the recorded signal. Remember, MPEG1 audio will work fine for you as long as you keep your levels high, but don't allow them to peak over 0dB. Set levels with a peak of around -10dB, carefully monitor them as you shoot, and you should be fine.

If your project involves live music, such as shooting a concert, it would be sensible to record the audio separately to ADAT, DAT, MiniDisc or another digital recording format. Take a mixed stereo feed from the mixing desk to your camera while on location. This is good practice regardless of the video format you are working with, as it will ensure the best quality audio for your final program.

If there is no live amplification of the audio, such as when shooting a small orchestral group, setup your own microphones and feed them into a small mixer which in turn is connected to your camera. Never rely on the camera's onboard mic for anything other than background ambience. Even a couple of PZM microphones strategically placed will sound infinitely better than using the camera's microphone.

Always use XLR Audio Connectors

You are probably familiar with the differences between balanced and unbalanced audio, so we won't dwell on that. If you're not, then it's important to get up to speed, because it matters! Boiling it down to the simplest terms, you should be using cables and microphones with XLR connectors. ¼" Jacks and RCA cables are sometimes okay for very short runs, but they have the tendency to pick up noise.

Lighting

One of the most often overlooked aspects of good video production is lighting. This is especially true with HDV, and it's even more critical if you are trying to obtain the much sought after 'film look'.

This book is not about lighting, which is a complex topic that deserves careful consideration by itself. If you believe it's a weak area for you (at the very least you should be familiar with three point

lighting), read as much as you can on the topic, and consider taking a course.

In general terms, lighting for HDV means two things: firstly, you need a sufficient amount of light to get the optimum image quality, and secondly, you need to 'craft' the light sources creatively so as to maximise the visual impact.

Taking the first aspect of having enough light, even if you are shooting outdoors, don't assume that you will have enough light. Many situations will require supplementary lighting to do justice to a scene. Remember that HDV cameras thrive on plenty of light.

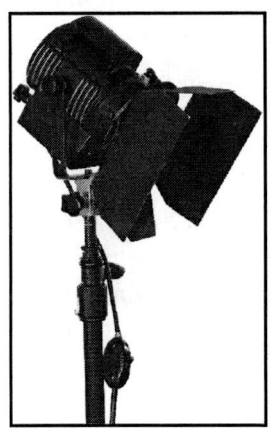

The second aspect of the creative use of light is more complicated and subjective. Supplementary light can be used to create atmosphere, give more of a '3D' or sculpted look to the two dimensional image, focus attention on a specific part of the frame, and much more. You can hopefully begin to see that it's well worth looking at lighting and devoting time to learn the art.

John Jackman's book, *'Lighting for Digital Video & Television'* is a recommended resource to get you going in the right direction.

Chapter 15
Media and More

Current models of HDV cameras accept only the small size Mini-DV or DVCAM tapes. In theory, any Mini-DV tape will record HDV, but in practice tapes should be selected with care, as you'll see. Your choice of media for HDV recording should involve much more than seeing who has the cheapest Mini-DV tapes.

One of the downsides to HDV recording is that the MPEG-2 compression increases the negative effects of tape dropout and errors. If a tape dropout occurs while shooting DV, only a single frame is usually affected whereas with HDV, the dropout lasts for an entire group of 15 frames (GOP), which is about half a second. So a dropout on an HDV recording is serious, it will be clearly seen.

Sony's Digital Master tapes are designed for use in HDV cameras. Image courtesy Sony Corporation Inc

To counteract this and to try and prevent dropouts occurring in the first place, key tape manufacturers have worked hard to develop new formulations that are manufactured to far more stringent standards.

Sony has developed a new tape formula for their Mini-DV and DVCAM tapes. Known as the Digital Master series, these tapes use two layers of active magnetic material and other enhancements to dramatically reduce the likelihood of dropouts occurring. According to Sony statistics, these tapes result in 95% fewer errors and 60% less dropouts compared to standard Mini-DV tapes.

Note that in the new range, Sony produce two Mini-DV tapes for HDV usage. One is the DVM63HD (in the consumer range), the other is DVM63DM (in the professional range). The only difference between these two tapes is that the DM version has a larger, more sturdy plastic case, similar to those that house DVCAM tapes. The physical tapes inside are identical. However, since the DVM63HD tapes are usually cheaper, you can save yourself some money by buying those (unless the larger plastic hard case is essential to you).

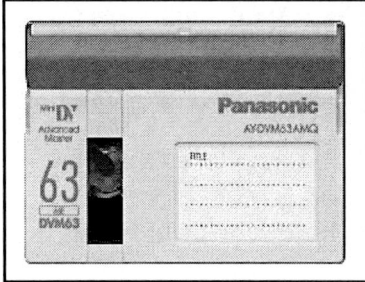

The Panasonic AMQ range is designed for HDV cameras
Image courtesy of Panasonic

Sony are not the only ones to have produced higher quality tapes designed for HDV. Panasonic have devised a new range referred to as 'AMQ'. As with the Sony Digital Master tapes, AMQ tapes have vastly improved magnetic density to reduce errors and dropouts.

Likewise, TDK recently entered the HDV media realm, with their HD DVC Media. The formulation features a dual-layer metal evaporation and diamond-like carbon technology. TDK claim that this helps to ensure lower error rates and fewer dropouts than their standard Mini-DV tapes. The tapes appear to be available in two formats-the HDV ProGrade Master which comes in a standard plastic case, and the HDV ProMaster which is packaged in a larger, more durable case.

Many cameramen ask whether it's worth spending the extra $8-12 on these specialist tapes. The answer to that is simple—it depends on how important your footage is. If you can live with

Media and More

the possibility of a half second dropout here and there then there is no reason why you should not use standard Mini-DV tapes. On the other hand, if you only have one chance at capturing your footage in any given situation, then the extra amount is a small price to play for peace of mind and less hassle in the edit.

Surely if you've paid around $4,000 upwards for an HDV camera, you want to get the best out of it?

To be fair, tape dropouts are not something that can be predicted. It's possible to shoot with standard Mini-DV tapes for many months without experiencing a single dropout. On the other hand, it's also quite possible that you could be plagued with numerous dropouts on your first recording. Again, it all depends on the importance of what you are shooting, so make your judgement on choice of tape stock accordingly.

One thing you should definitely avoid is re-using tapes that you have already recorded onto. With HDV that is asking for trouble. These days tape stock is so cheap in the grand scheme of things that it's false economy to reuse tapes, and that applies to any camcorder.

Something else to note while we are on the subject of tape stock is that once you choose a brand of tape it is best to stay with that brand every time you buy tapes. Different manufacturers use different formulations in their tapes, and it has been reported that mixing brands of tapes can cause a negative interaction, to the point where the recording heads of a camcorder can get gummed up and cease to function as they should.

If you have no option but to use different brands of tapes because you have them in stock, be sure to run a head cleaning tape through the camera in between using the different brands.

How to Shoot, Edit and Distribute HDV

Something else that seems to cause confusion is the matter of 'blacking' tapes. Is it essential to black a tape by recording with the lens cap on through the length of a tape in order to lay a continuous timecode track onto the tape? In short, no. This used to be common practice when using analog cameras and decks because the editing decks would have trouble locating timecode if it was broken between scenes. Modern digital cameras and decks do not have this problem, even if you start and stop recording numerous times for different scenes.

If you already have a tape supplier and you're happy with them, that's fine. However, I would like to recommend Edgewise Media to you. I want to be honest in telling you that I do have an affiliation with Edgewise in that I write their weekly newsletter. However, I have no hesitation in recommending them because I also provide services for other companies in our industry, some of whom I would not recommend to you.

Edgewise are one of the few truly legitimate, approved distributors by the major manufacturers, they always have plenty of stock on hand in their multiple warehouses around the country, and they provide outstanding customer service, probably the best in the industry actually. You can find them at: www.edgewise-media.com or call 1-800-959-5156. Please tell them I sent you, and they will look after you.

Chapter 16
How to Get the 'Film Look'

This was a difficult chapter to find a home for within this book because it covers all aspects of the production process, including shooting, editing, and distribution.

In recent years there has been an intense interest among many videographers to imitate the look of film, using video as the recording and playback medium. By film, we mean any camera that uses reels of film, including 16mm, 35mm and even the IMAX format.

Others within the industry are interested not only in achieving the film look with video, but they also want to release their final project on film, to be viewed in theaters. George Lucas famously went down this route with his most recent Star Wars movies. They were shot in HD, post-produced on computers and output to film as the last step. Regardless of your opinion of Star Wars as a movie series, no one can question that technically they look magnificent.

What's interesting about achieving the film look is that, from a technical point of view, it's actually an odd thing to do, because the many aspects of video that are technically *superior* to film are the very same things that video producers don't like. For example, consider the following three comparisons:

1. Film is inherently grainy and soft, whereas video is 'clean' with razor sharp definition.

2. Video compares more closely to what our eyes see—it's realistic, whereas film captures a more artistic look.

3. Video has a higher frame rate, namely 30fps for NTSC, which produces a smoother motion, especially since it is interlaced. Film's 24fps progressive frame rate produces a somewhat jerky, semi-blurred motion by comparison.

So what we are saying is that psychologically, to many viewers the flaws, softness, grain, jerky motion, and artistic style of film as a whole is more pleasing to the eye compared with video. Clearly, the viewer's perception has as much to do with psychology as technology.

Of course, the film versus video debate is somewhat subjective, and to a large degree it depends on the subject of the piece. For example, many people would agree that film lends itself to more artistic and dramatic subjects whereas video seems to lend itself better to documentaries and factual topics where there is less focus on influencing emotion.

Thinking of Releasing Your Video as a Film?

Most of us at some point have had dreams of releasing our masterpiece on film to view it in all its glory in a cinema, perhaps even having it released around the country or even worldwide.

Let's be realistic here. The costs of making even one master film print from a video project will be prohibitively expensive for most people. It's a VERY expensive business (at least $300/minute to transfer video to 35mm film). Unless you have serious financial backing, you would be wise to set aside your dream until you can afford to plough tens of thousands of dollars into it. Every year, countless budding filmmakers go broke pursuing this dream because they did not count the cost before they started.

How to Get The 'Film Look'

Whatever your view on the video vs film debate, it's good to be aware of the differences between film and video in order to know how best to approach your own projects.

Don't be influenced by over-zealous industry peers who tell you that shooting for the film look is essential to your success. Quite frankly, it's not. It's a tool that may be suitable in some circumstances, so always try to look at it from that standpoint.

With that caution, let's consider how you can achieve the film look with HDV.

The Shooting Phase

If you want to obtain a film look to your video projects while continuing to distribute them in a video format, give careful consideration to the following seven 'tweaks':

1. Shoot in the 16:9 Aspect Ratio

The widescreen format immediately conveys the idea of a movie to viewers. Think about it. Not only are people used to seeing movies at the theater in the widescreen format, but they know that many films shown on 4:3 TV's are formatted in the letterbox style, with black bars at the top and bottom of the screen to maintain the original composition.

This of course is the easiest of all 'tweaks' because HDV is inherently 16:9 widescreen. In fact, this tweak has already been done for you!

2. Frame Rate

Since film is recorded at 24fps, the ideal approach with video is to use a recording system that is optimised for 24fps progressive.

For HDV, the only camera that currently does that is the JVC HD-100U.

If your camera does not allow you to record 24fps progressive, the next best option would be to record 25fps, because 25fps is obviously much closer to 24fps than 30fps is. The Sony Z1 can be set to record 25fps.

Many other HDV camcorders have modes that aim to imitate the look of film. This would be the least favored option but it's something to consider if there is no other choice. On the Sony FX1 and Z1, the 'CineFrame' modes process the image to make it look as though a frame rate of 24fps was used, even though they do not record true 24fps progressive. The Canon XL H1 similarly features a 24f mode to imitate the look of 24fps.

Ultimately, it's a matter of taste. Many people are happy with the Sony and Canon 'pseudo' film modes, whereas purists will say it has to be true 24fps or they won't touch it.

3. Gamma Curve

It's good to remind ourselves that film and video each capture images in very different ways. Film is a very smooth, organic process where the images are acquired chemically on light sensitive material, whereas video is a rigid record of digital data with much less shading. For that reason, film has a much wider gray scale than video, as well as the ability to show a smoother range of colors, shadows, and highlights. This is what is known as the gamma curve.

The difference between video and film gamma curves is most noticeable at the extremes of the ranges. When viewed on a test instrument, video has a very straight gamma curve, whereas film has an 's' shaped curve where the extremes have a slight, but noticeable curve.

Translated into something you can relate to, film is able to show a wide range of gray on the way down to black, whereas video sharply drops off from gray to black in a few sharp steps. Modern cameras have become far more adept at imitating film than they used to, and many HDV cameras now allow you to adjust the gamma settings to create more of an 's' curve. For example, in the Sony FX1 and Z1, Sony has helpfully provided two presets called 'CinemaTone', of which the second preset seems to be the most useful.

4. Depth of Field

Depth of field is the ability to keep the foreground subject sharply in focus while the background is blurred. This focuses the viewer's attention on the subject in a subtle, but powerful way, and it's a look that is synonymous with film.

With film, the film itself as well as the lenses used, allow for easy manipulation of depth of field. With video, it's far more difficult to do this because camera CCD sensors are so small compared to 35mm film frames, which is one of the reasons why it's unusual to see scenes on video that were recorded with a narrow depth of field.

One way to achieve a narrow depth of field is to position the camera further away from the subject and zoom in. This has the effect of making the distance between the foreground and background seem further apart.

With depth of field, the foreground subject is sharply in focus while the background is blurred. Image courtesy of Red Rock Microsystems LLC

In addition, one or more neutral density filters can be used to reduce the amount of light coming into the lens, which in turn widens the aperture of the lens, which also helps reduce the depth of field. Remember, the wider the aperture, the shallower the depth of field.

Another alternative is to use an add-on device such as RedRock Micro's 35mm adaptor, which allows you to use 35mm lenses on your HDV camera. With such lenses it's far easier to adjust the focus to achieve the desired look. There's more information on the RedRock Micro adaptor later in the book.

5. Use Filters to Soften Edges

We've already discussed the advantage of using a neutral density filter as an aid to gain depth of field. Another filter to consider using is a Pro-mist. These filters are available with different levels of the effect, so to varying degrees they reduce contrast and soften edges, again imitating the look of film. As an alternative, most cameras allow you to directly reduce the edge sharpening 'enhancement' that is a hallmark of video. You may find that tweaking that setting will soften the video enough to eliminate the need for a filter. Another benefit of toning down edges is that it reduces the likelihood and severity of motion or compression artefacts.

When using filters this way to soften the image, be sure to evaluate each shot independently, as different scenes will likely require lesser or greater amounts of the effect to be applied.

6. Support the Camera

Handheld shots are a definite no-no when endeavouring to create a film look. Yes, 'shaky' footage is sometimes used as a tool in movies, but it's handled within very careful parameters, and what is perceived as handheld footage is often a gentle controlled motion of a specialised supporting device. Unless there is a very powerful reason not to for certain shots, use a tripod, steadicam, crane or other supporting device at all times. Your footage should be rock solid.

7. Lighting

Lighting for a 'film look' involves creating a modelled look with lots of shadows, rather than the flat-looking wash or flood of light that is so common with video.

Also, bear in mind that film has a much higher dynamic range than video, so you may need to use some fill lighting to show details in dark areas of a scene. Aim to avoid seeing any large areas of black or white where there is no detail.

Lighting for film is far more artistic than lighting for video. Don't be afraid to experiment with different light sources placed creatively to 'sculpt' the scene.

Now we come to the editing and post-production phase. The following techniques should be considered as aids in achieving the film look with HDV:

1. Convert Footage to 24p

Since the DVD format allows you to display 24p, that's the closest video can come to the frame rate of film. Unless you are able to shoot in 24p, you will need to process your footage. After the HDV shots have been captured, you need to de-interlace them to make them progressive as well as to change the frame rate to 24p.

The DV Film Maker software from www.dvfilm.com can convert any HDV footage to 24p. The advantage to this software compared with many other solutions is that it analyses each frame to determine where there is movement, and only those pixels are affected. This results in higher quality images and smooth motion. It's also a standalone utility, so it's not dependent on any other software or hardware.

Another solution worth considering is the CineForm software. Aspect HD has the ability to convert imported HDV to 24 fps on the fly. You can then use your editing software to de-interlace the footage to make it progressive. Depending on what solution you use to handle the video conversion, bear in mind that the audio also needs attention. From 25fps, it will need to be slowed by 4% to make it useable with 24p video.

2. Add Film Grain

The very nature of film means that it is grainy. The amount of grain depends on a number of factors, including how old the film is supposed to be, and of course a similar effect can be imitated in the editing stage. Unless you deliberately want your film to look very old, keep the applied effect very subtle. If you watch modern films you'll see that the grain is far less pronounced today than it was in the past, the reason being that manufacturers have made improvements to film stock to decrease the amount of visible grain.

If you plan to actually output to film, it is essential to do some tests to see what effects work, and in what 'dose'. Depending on which editing system you use, there are plug-ins available that will enable you to accomplish this process relatively easily, bearing in mind that you will have to wait for it to render of course.

3. Add Film Artifacts

Film artefacts are the anomalies commonly associated with older types of film, such as dust, scratches, and hair. Software such as Adobe After Effects and Red Giant's Magic Bullet do a good job of imitating these anomalies. Other products are referenced at the end of this book in the additional resources section.

As with the comments about film grain, artefacts should be added according to the type of film and degree of agedness that you want to convey. The older the film is supposed to be, the more film artefacts it is likely to have.

4. Add Film Colorization

Many film directors like to choose film stock for its particular color characteristics, and they often apply further colorization in post. They do this to give the film a uniform look, as well as to help convey subtle clues about where the movie is set, the emotions involved, and so on.

You may have noticed that this is even being done with many popular TV shows too. For example, the TV show CSI: Miami has a predominantly warm look to it, conveying the sun, dry and dusty geography, which in turn reinforces the natural view that most people have of that region. Contrast that to the predominant green tinge that is used on the original CSI series, set in Las Vegas.

Many of the same software plug-ins that were described previously are also adept at colorization. However, a leader in the field that's also within the budget of most readers has to be Red Giant's Magic Bullet suite. Their software enables you to choose or adapt presets, many of which are given names that were inspired by popular movies such as *The Matrix*, *Three Kings*, and *Traffic*. You can download a free trial from their website.

Chapter 17
How to Edit HDV

Do you want the good news or the bad news? Well, let's start with the good news. It is now possible to edit HDV on a sub $2000 computer. The bad news is that certain aspects of the editing workflow currently slow to a snail-like pace. For example, there is no cost effective and speedy way to render and output a finished HDV project, and then there is the hurdle of few practical choices for distribution.

It's now possible to edit on a relatively inexpensive computer

Early adopters who have been struggling with these issues are eagerly hoping for a resolution, sooner rather than later. Right now it's very much like going back to the early days of DV.

There are a number of quirks with HDV that make it quite different to DV. For example, HDV devices output what is known as a 'transport stream' down the Firewire cable. If you've used the Beta SX format you will probably be familiar with transport streams.

A captured transport stream has the file extension .ts or .m2t, but as the name implies, it is a format that was designed purely for transporting the signal. It's not designed to be edited, even though some software manufacturers have gone down that route. The alternative editing process involves capturing the transport stream from the camera first, and then converting it to another format for the edit. We'll discuss that in a moment.

Very surprisingly, many of the major manufacturers of editing hardware and software left HDV users high and dry—with no practical solution for editing until fairly recently, despite the fact that HDV cameras had already been on the market for several years. Why did they wait so long? If they were waiting to see if the format caught on, surely the fact that in 2003 no less than four major manufacturers announced their full support of the format should have been a good indication that HDV would be around for awhile? We mere mortals are obviously missing something…

The bottom line at this stage is that the technology for post-producing an HDV project is advancing, albeit more slowly than end users of the technology would like. Yes, you can work with the format in a relatively stable environment, but it's not yet a painless process for the average editor.

MPEG-2 Compression has a lot to Answer for!

The major challenge with HDV editing is the MPEG-2 compression that the format uses. It's something of a mixed blessing. On one hand it allows the HDV format to exist, yet on the other hand it presents a serious challenge to work with.

Compression in itself is not the problem. For example, DV is compressed, and each frame stores the complete pixel information to be able to edit at any frame. This is known as intra-frame compression. HDV on the other hand uses a different method, known as inter-frame compression, which is where the challenge lies.

You already know that HDV has to cram a lot more pixel information into the same amount of Mini-DV tape because of the much larger frame size, so how does it do it? Well, it has to discard much of the data and fill in the resulting 'gaps' by referencing back and forth to the few complete frames that are recorded.

With MPEG-2, there are three types of frames, denoted by the letters I, B, & P. Only 'I' frames contain the complete pixel information for the entire frame—every pixel in the frame is recorded. However, only one of these 'I' frames occurs every 15 frames (two each second), so you can start to see the challenge with piecing together the remaining incomplete frames.

'B' frames (bi-directional) compare the frame before and after the current frame for changes, and 'P' frames (predictive) compare the current frame with just the previous. This combination of the three types of frames is called a 'group of pictures', or 'GOP' for short, and there are variations of GOP structure for different types of MPEG encoding.

Let's simplify all of this with a basic description of what happens with MPEG-2 compression. Imagine you were capturing footage of a man walking in front of his house. The first frame—the 'I' frame—contains every single pixel in the image, it's the perfect image. In the next frame, as the man starts to move across the screen, most of the image (in this case the house), will remain the same. The MPEG-2 process intelligently recognizes that a large number of pixels in the second frame are exactly the same as the previous frame. Instead of recording every pixel individually again, it simply notes 'to and from' reference points where pixel data is the same.

There's no doubt that HDV MPEG-2 is a clever system of compression, but you can begin to see the downside. Anytime a computer looks at MPEG-2 video, it has to decipher and piece together all those individual frames that don't have the full pixel information. That requires enormous processing power as well as fast access to and from the hard disk.

How to Edit HDV

Don't worry if all this sounds too complicated. It's not essential to know the intricate details about the MPEG format. The important thing is that you understand in general terms why HDV is more difficult to edit than DV, because that knowledge will help you to make the right choices.

Frustrations with MPEG-2...

Currently, no editing solution offers frame-accurate capture for HDV. All you can do is start the capture at roughly the right point and then stop it at an appropriate point at the end of the scene you want. Or you can capture an entire tape in one go, but be aware that the capture program might automatically divide the footage into smaller files, naming them with an extension (such as clip1-001.avi, clip1-002.avi etc) to indicate they are part of one long capture. Some capture utilities are able to detect scene changes and name a new clip automatically at the start of each new scene, which is a great feature to look for.

While we're on the subject of things you can't do with HDV, when you capture footage from a camera or deck, you will not be able to see a preview of the video on your computer monitor. The only way to tell where you are on the tape is by looking at the LCD monitor on the camera or deck, as there is no real-time preview.

Also, when you output your final HDV project to tape via Firewire, there is no way to do an insert edit, because of the complex GOP structure of MPEG-2 video.

Where is HDV technology at today? What about the major issue of outputting a finished HDV project?

The support for HDV is slowly improving, and it's gaining momentum. Several hardware solutions now offer component

How to Shoot, Edit and Distribute HDV

video output in real-time, enabling you to record to an HD capable deck that accepts component inputs at the appropriate resolution. However, this still does not solve the requirement that many users have—to quickly and easily output via the Firewire cable back to an HDV camera or deck.

Some editing solutions preview the timeline in real-time—at standard definition resolution. While that might be adequate if your final output is going to be standard definition video, surely it is counter productive if you are trying to edit an HDV project? How are you supposed to check how your video footage *really* looks?

HDV that's been down-converted to DV looks VERY different to native HDV. For example, an image that is out of focus when viewed in HDV on a high resolution monitor can look in focus when down-converted to DV (standard definition video). Likewise, if you set-up a chromakey effect and view the result, it might look great in standard definition, but be in need of significant adjustments to be accurate on an HD monitor.

You may be thinking that the answer to all these issues is to rush out and buy a whopping dual processor beast of a computer? Sure, that will help, but it's not the cure-all solution you might think it to be. You see, the MPEG-2 compression used for HDV is VERY complex, and currently even the fastest computers loaded up with as much RAM as you can cram in are still not up to the task of outputting an HDV project in real-time back to an HDV deck or camera. There's more to it. To work with MPEG-2 video effectively (in real-time) requires a dedicated processing chip on a graphics card in order to decode and encode the video. Some manufacturers appear to be working on such a solution.

If you've had experience in outputting a *DV* project to DVD then you'll know even that is a time consuming process. To convert

How to Edit HDV

any video to MPEG-2 (the format that DVD uses) is time consuming and CPU intensive. Multiply that by at least four times and add a few other complications into the mix, and that's what you are dealing with when outputting HDV.

To reiterate the point made earlier, it's still early days for HDV editing. The technology has a long way to go before it settles down to the same level of stability and ease of use that we have become used to with DV. Add to that a mix of manufacturers who offer vastly different approaches to editing HDV, yet who all claim to have the best solution! However, don't be put off. You *can* produce an HDV project from start to finish, right now, and that's what counts.

Transferring HDV To / From Your PC

No doubt you are already familiar with transferring data from your DV camcorder to your computer via a Firewire cable, a technology which is also known as IEEE1394, iLink or OHCI—depending on who's talking about it. As a quick refresher,

A 6 pin to 4 pin Firewire cable

Firewire connections come in two sizes: 4 pin and 6 pin. The smaller size 4 pin connector is generally used on camcorders and laptops, whereas the larger 6 pin size is often used on desktop computers and some decks.

The exact same physical cables and connections are used with HDV, so that is good news. However, that's where the similarity ends. Think of a Firewire cable as a pipe through which a variety of liquids could flow. With HDV, the type of data that flows is very different to DV. Furthermore, the two types of data are not compatible, which means that you cannot capture HDV using the regular DV settings in your capture or editing software.

86

Three Editing Options

Currently, the marketplace offers three main ways of editing HDV. Each one has pros and cons, and the fact that manufacturers are each vigorously promoting what they consider to be the best way of editing HDV, demonstrates that the technology is still in a state of flux. They can't all be right can they?

Interestingly, a few manufacturers are offering a combination of two of the three options within a single system, and letting users choose which one they prefer.

Let's review the three options:

1. Native HDV MPEG-2 Editing

Native HDV editing manipulates the MPEG-2 video directly, even though transport streams were never designed to be edited. This method is by far the most processor intensive because the computer has to decode the MPEG-2 streams, and the hard drives have to work hard to process the data that is flying back and forth. Unless you have a very simple, cuts-only project, native editing is not a good idea at the present time.

The only redeeming factor of native editing is that there is little quality loss in the process. Some manufacturers claim there is no loss at all, but this is not strictly true. A conversion happens when changing the captured 'transport stream' to a 'program stream' for editing, and then back again when exporting the final project; however, the losses are slight, and the process happens far more quickly than the alternatives.

2. Proxy Editing

Proxy editing uses lower quality 'substitute' clips until the final output stage, at which time the full resolution HDV clips are referenced and used to create the final render. This method is currently the least favored overall because it adds another layer of complexity to the edit, and as such it is only offered by a couple of software companies.

3. Intermediary Format Editing

Editing with an intermediary codec is currently the most popular choice for HDV users mainly because it provides for responsive timeline functionality even with relatively 'low powered' computer hardware.

How does this system work? Immediately after the video is captured, software converts (transcodes) the video into an intermediary format, usually an AVI file, with a codec that's easy to work with. Clever compression algorithms enable the editor to work with a quality of video that looks the same as HDV, without the problems associated with working with MPEG-2. That's a major advantage and it opens up more creative capabilities that can be handled in real-time.

One disadvantage is that transcoded files are much larger than their MPEG-2 counterparts, so they take up significantly more disk space. However, with hard disk space relatively cheap these days, that should not present a serious problem.

Once the editing is complete, the project needs to be transcoded back to another format, such as MPEG-2, which can take considerable time (depending on the length of the project), and although in real terms no quality loss is visible, there is inevitably some quality loss in the process.
That's the three editing options. In the next chapter we'll examine all of the currently available editing applications that use these methods.

How to Shoot, Edit and Distribute HDV

'Real-time' Editing

Sadly 'real-time' has become a widely misapplied term in the video editing industry. Some software and hardware providers take substantial liberties with what they label as real-time. Beware! For example, any editor who could only play back their timeline at a fraction of the full resolution, with regular stuttering, would hardly call that real-time. Yet some manufacturers do.

The bottom line: when considering an editing solution, always question what the *actual* performance of the system is, regardless of any marketing claims that are made. The best approach is to see a demonstration of the application you are considering, taking careful note of what kind of computer power is running behind it. If there is a free trial that you can load on your computer, be sure to take advantage of that, too.

Let's now consider the possible workflows. There are three ways of working with an HDV camera:

1. Shoot DV or DVCAM, Edit DV or DVCAM

2. Shoot HDV, Edit DV

3. Shoot HDV, Edit HDV

Let's take each of these in turn. Firstly, why would anyone buy an HDV camera and then choose to shoot and edit in DV or DVCAM mode? Well, to be fair, there may be some situations where HDV would be overkill. For example, if you were shooting a seminar that needed to be good quality but where HDV would be unnecessary. Yes, you could shoot HDV and down-convert, but focusing is more difficult with HDV, and when dropouts occur they are more serious. Shooting DV or DVCAM in such a situation would be easier and a sensible choice.

The second option is to shoot HDV, then edit in DV mode. Footage that's been down-converted from HDV to DV produces better quality DVD's than footage that was shot in DV, so that's a good reason to use this workflow.

Another reason to use this approach is when filming a wedding or any shoot when there is even a remote chance that the higher resolution HDV footage could be valuable in the future. By shooting HDV you can deliver a standard definition version now, store the tapes, and produce an HDV version down the line, perhaps when suitable HDV delivery methods have become widespread. You can also charge the client twice for working on the same project—when you deliver the SD version and again when providing the HDV version. Many videographers are setting themselves up with the potential for a good source of future income by doing this.

The final workflow option is probably the most logical for many—shoot HDV, and edit HDV. Of course, not everyone has the capabilities to edit and distribute HDV, but this workflow obviously provides the best quality.

Using DV in an HDV Project

On occasion you might need to incorporate some DV footage into an HDV project. The challenge is the substantial difference in frame size. However tempting it might be, don't try and scale your DV clip(s) to HDV resolution, the results will never be acceptable unless you have access to specialised equipment for that purpose. Instead, try and creatively incorporate the footage by playing it in a 'window', in other words inside the HDV frame. If you've viewed an IMAX film, you'll have seen this technique used. To incorporate an old film clip or a piece of video that is impossible to acquire with an IMAX camera they simply play the segment in a window within the rest of the black IMAX frame, and it overcomes the quality issue.

A further way to draw attention away from the lower resolution of the SD video clip is to place the clip over an HD resolution animated graphic background, such as Digital Juice 'jump backs'. As a side note, all HDV capable editing systems will also edit DV only projects, you don't need a separate system for DV.

Capturing Component Video

Although most editors will probably capture via Firewire, the option exists to capture the component signal, which of course is very high quality. Companies such as Black Magic produce graphics cards that allow you to capture uncompressed component video at full HDV resolution. The downside to this approach is the enormous file sizes that result.

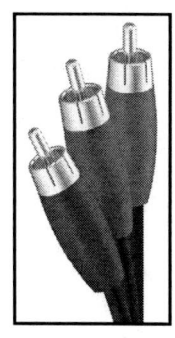

Component video is comprised of three cables—red, blue and green. The audio is conducted separately.

Firewire 800

Firewire 800 is to Firewire what USB-2 is to USB. It's the next progression in speed. At the time of writing, few Firewire (FW-800) capable devices existed, although support for the technology is gradually gaining momentum.

Although hard drive speeds are not the most important factor when editing HDV, they are still important. One area where FW-800 opens up some possibilities for the HDV editor is in external hard disk storage for your captured footage. Companies such as Lacie produce some excellent external RAID arrays in aluminium enclosures that give you the advantage of portability and the ability to connect your capture footage to different computers.

Of course, to use any external FW-800 device, you would need a FW-800 port on your computer. Thankfully, the FW-800 connector is a new design so as not to cause confusion with the original Firewire cables. PCI FW-800 cards with a combination of FW-400 and FW-800 connectors are a neat solution for this purpose. If you're not up to speed with computer technology to the extent where you can build your own system, at least you'll know what to ask for with your next upgrade.

> **Planning to Output to Film or HDCAM?**
>
> If you plan to export your final project to a higher quality format such as film or HDCAM, make the change to the highest quality digital format for editing as early as possible in the production chain. In other words, up-sample your HDV video before editing.

Editing Options Summary

Weighing up all factors, editing with an intermediary codec is probably the best option at this time for most people. The main factors influencing that statement are:

1. The substantially lower computer hardware costs involved is a major factor—after all, the essence of HDV is about working within a tighter budget.

2. The enhanced timeline responsiveness and ease of use that is obtained as a result of working with an intermediary format.

When Dual-Core, dual processors become substantially cheaper and native editing becomes as responsive as working with DV, the balance will probably swing the other way. As it stands, in every test that I have witnessed, even the fastest computer setup is still not as responsive when editing native video, compared with editing using an intermediary codec.

Let's now review what manufacturers currently offer you in terms of HDV editing hardware and software.

Chapter 18
Editing Hardware & Software

In this chapter we will look at the major editing options for HDV that are currently available, presented in no particular order.

Main Concept

Main Concept has produced a plug-in for Adobe Premiere Pro that allows you to work with native HDV files. However, as you might expect, the timeline is decidedly sluggish even on a well specified computer, and the only advantage with the Main Concept solution seems to be that outputting to tape is fairly quick because the project stays in the native MPEG-2 domain throughout.

The software allows you to see a real-time (including scrubbing) standard definition output of the timeline on your HDV camcorder, via the Firewire cable from your computer. To see a larger image you could attach a monitor with a larger screen to the composite, S-Video or component output of your camcorder.

Ulead Media Studio Pro

Ulead's flagship video editing software, Media Studio Pro has chosen to offer two workflows for HDV—you can edit either native MPEG-2 video, or in a proxy mode called 'smart proxy'. As noted earlier, the latter is a compromise to overcome the challenges of editing HDV files directly, especially on the average powered computers that are typically in use by many editors.

In the 'smart proxy' mode, MSP works in the background to create lower resolution copies of the clips from your project.

Editing Hardware & Software

When it comes time for the final output, the 'links' refer back to the original, full resolution files to create the finished video. Another application of the proxy system would be to keep the original high-resolution files on a central server, and share proxy files amongst multiple editors simultaneously. Each one could work on a section of a project and then combine their efforts at the final stage, when the original footage is referenced and compiled.

On the plus side, only sections of footage that are changed during editing need to be rendered. This saves considerable time at the output stage. Also, quality loss would be minimal.

MSP does allow you to mix different file formats on the timeline, so it will handle all types of MPEG, DV and HDV within one project.

Ulead Video Studio 9

Although it's not a professional editing application, in the latter part of 2005 Ulead provided an HDV plug-in for their *Video Studio 9* software. It enables consumers of Sony's HDR-HC1 HDV camcorder to capture and edit native 1080i HDV, and then output to WMV-HD or DVD. By offering this solution, Ulead probably captured a significant share of early adopters of Sony's HC1 camera who were looking for a way to edit their home videos.

CineForm

CineForm have rapidly established themselves as the foremost provider of HDV plug-ins for popular editing software. When you consider that both Sony and Adobe chose CineForm technology rather than develop their own solution for editing HDV, that in itself speaks volumes.

An entirely software driven solution, CineForm products transcode (convert) captured HDV to create 'CFHD' files

How to Shoot, Edit and Distribute HDV

(CineForm High Def), using CineForm's special wavelet codec. The resulting CFHD files are AVI's, for compatibility with multiple Microsoft Windows applications, which opens up far more editing possibilities than .ts or .m2t files.

CFHD files are around four times larger than native HDV files (you'll need lots of hard disk space), but are 'visually lossless' when compared with the original HDV files. That's a bold claim, but it proved true in my own tests. As a guide, an hour of 1080i HDV CFHD files will consume about 40GB of hard disk space.

The transcoding process takes place at the same time as capturing, but depending on the speed of your computer it might take a little while after the capturing has finished to complete the transcoding. Time to put the kettle on...

CineForm offers three levels of functionality in the form of their three software products: Connect HD, Aspect HD, and Prospect HD.

Connect HD

Although Sony Vegas already includes CineForm technology as its core HDV engine, Connect HD is designed to enhance the functionality of Vegas further. It gives better responsiveness on the timeline, faster processing, and allows basic editing and effects in real-time.

Aspect HD

Aspect HD, is primarily designed to be used with Adobe Premiere Pro or Adobe After Effects. As with Connect HD's enhancements of Sony Vegas, it extends the functionality of the standard HDV engine in Premiere Pro. In

95

Editing Hardware & Software

fact, CineForm claim to quadruple the performance of the basic Adobe Premiere Pro HDV functionality.

Aspect HD allows multiple layers of HDV to be edited, along with numerous real-time effects, motion titles, color adjustments, dissolves, wipes, page peels, and picture-in-picture effects. Granted, the effects are not as diverse or numerous as the complete range found within Premiere Pro normally, but considering they work in real-time, it's a very usable solution right now.

A further benefit of the CineForm technology is that Aspect HD fully supports capture and editing of the CineFrame modes of the Sony FX1 and Z1 camcorders, as well as the option to output 24p from the timeline. CineForm technology was the first to offer these benefits.

Something else that was touched on earlier but is worthy of repeating, is that all CFHD files are fully compatible with Adobe After Effects, which opens up some interesting options for compositing HDV.

Aspect HD will undoubtedly appeal to many HDV editors, and it's currently one of the most widely used solutions in the current marketplace at a very competitive price point.

Prospect HD

Prospect HD is the top of the range solution from CineForm and can only be used in conjunction with certified hardware. Primarily, it's designed for high-end HD editing; its HDI-SDI interface providing real-time monitoring of the timeline as well as export to tape for output. HD-SDI and SDI in and out can be implemented using an AJA Xena-HS card. As such, it is over specified for most HDV users.

A capture application—'HDLink'—is also included with all of the CineForm software options; it's a thoughtful addition, although you may find that your editing software provides more functionality. When capturing, a choice of three quality options are offered: low, medium, and high. The default setting—medium—is the best choice for retaining quality without using the substantial additional amount of hard drive space that the high setting requires.

Adobe Premiere Pro

As mentioned previously, Adobe adopted CineForm technology to provide basic HDV capturing and editing functionality within Premiere Pro 1.5. Further functionality requires the use of other suitable plug-ins—CineForm's Aspect HD is highly recommended.

Avid Xpress Pro HD

Some might say that Avid, a name synonymous with video editing, were slow on the uptake with HDV. However, when they started to support HDV in the latter part of 2005, the result—in the form of Avid Xpress Pro HD—was worth the wait. Choosing to offer a native editing solution, the software offers some real-time functionality on a very fast computer.

Avid states that many of their users want to mix HD, HDV, and DV on the timeline, so they have heavily promoted that functionality of their software. It really works, too. Any format of HD, HDV, or DV can be imported straight onto the timeline. That's pretty spectacular when you consider the vast differences between, for example, DVCPRO HD, XDCAM, and HDV.

Another feature of Avid's 'Open Technology' is the ability to do multi-cam editing. If you're not familiar with the multi-cam concept, it allows you to cut between multiple cameras (layers of video) while viewing them all at the same time. Picture the control room of a TV station, where the vision mixer looks at a row of monitors and follows instructions from the Director as to which camera should be live at any given time. Multi-cam editing follows a similar concept, with the difference that the multiple camera views have already been recorded to tape and then captured. Again, the multi-cam editing option can use a mix of file formats.

The main way Avid achieves acceptable performance when editing HD is by their intermediary codec, named 'DNxHD'. This advanced compression allows high quality HD to be edited and delivered within the same bandwidth and storage space as DV.

Canopus Edius NX & Edius Broadcast

Canopus has built up a good reputation in the video editing arena, and they were fairly quick to see the potential of HDV. As a result, they had a substantial head start in launching a hardware/software combination solution for editing HDV in the form of 'Edius NX'.

The Canopus' Edius software does not require the associated hardware card for editing HDV—you can still use Edius to edit HDV without it. The trade-off is much slower responsiveness of course.

The key positive aspects of the Canopus hardware are that it provides full resolution component HD output while editing, as well as accelerated MPEG-2 functionality for certain tasks. The downside is that it requires a *very* fast, dual processor computer to function.

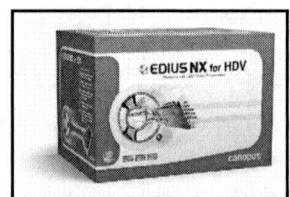
Image courtesy of Canopus

How to Shoot, Edit and Distribute HDV

Edius can work with HDV in two ways—to capture HDV natively as .m2t MPEG transport streams, or to transcode them to the Canopus HQ editing codec. As with the CineForm solution, you will usually need to wait for the transcoding process to finish after the capturing has stopped. As you might expect, working with the HQ transcoded files offers much faster performance on the timeline compared with native editing, yet for all intents and purposes it looks the same as the native footage. Native editing still has a way to go before it can compare in this regard, even with the NX hardware.

Canopus makes bold claims about the image quality of their HQ Codec, in particular stating that they have superior Luma and Chroma sampling of HD video. Honestly, it's very difficult to see any differences compared to other HDV editing systems, when monitoring on an HD capable screen. However, the adjustable bit-rate feature is a good idea and will be handy for some editors, as it allows you to select the highest bit-rate that works with your level of computer hardware. That means the output quality can be improved if you are using a high-end computer.

As for adding effects and filters when using transcoded footage, Edius NX allows you to chromakey, blur or sharpen, color correct and slow-mo the footage, all in real-time, even with several layers of HDV on the timeline—very impressive. By contrast, if you are using native HDV clips (MPEG-2 TS), only one layer of video can be edited in real-time on an average power computer (Intel P4 with a 2.8Chz processor).

A major downside to the current Edius NX system is that it does not allow you to monitor 720p in real-time. This might change with a future release, but for now users of JVC cameras will have to scale their 720p footage up to 1080i if they want to use the Canopus system.

Editing Hardware & Software

As a side note, although the hardware is primarily designed to work with Canopus' proprietary editing software, *'Edius'*, Canopus claims that it will also function reasonably well with Adobe Premiere Pro. However, to be honest, with what some might refer to as "half-hearted" support for Premiere Pro, for system stability you are better off with the CineForm solution for editing HDV in Premiere Pro.

At the end of 2005, Thomson acquired a sizeable share in Canopus with the intention of boosting its own Grass Valley range, and one of the early developments was the launch of *Edius Broadcast*, a software bundle that integrated Edius Pro 3 with Edius Speed Encoder for HDV, which claims to harness the power of Dual-Core computer systems to export HDV MPEG-2 at faster speeds than were previously available.

Matrox Axio HD

Matrox is another company that appears to have upset many editors by being slow to offer a dedicated solution for HDV. Their DV editing hardware, the RTX.100, has been one of the best selling solutions in the marketplace over the last few years, so the lack of cost-effective support for HDV is puzzling to say the least.

In fairness to Matrox, they do have a platform that will work well with HDV, but at a price—a much higher price than HDV specific products in the marketplace. That platform is Axio HD.

As the name suggests, Axio HD was primarily designed to be an HD editor, and the HDV support is a feature that was added with the release of version 1.5. Axio HD, which uses Adobe Premiere Pro as its interface, has excellent real-time capabilities thanks to the dedicated hardware involved. And it is true real-time, unlike so many claimed real-time systems that turn out to be 'slow-time'. You can throw multiple layers of HDV video and multiple layers of graphics on the timeline, add effects, and the program will playback in real-time with no rendering required. It's very impressive, but unfortunately as mentioned earlier, it's at a price point that will be beyond the affordability of most HDV editors.

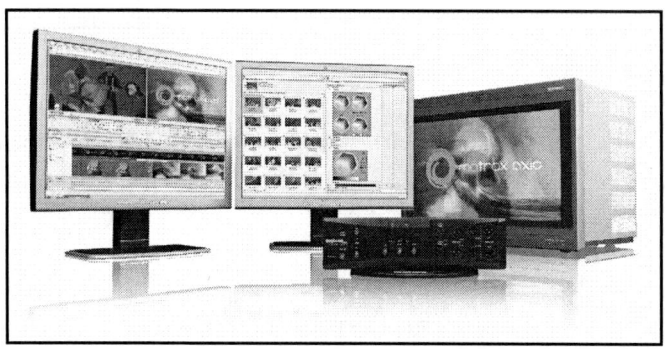

Axio makes easy work of HDV, at a price.
Image courtesy of Matrox

As you might expect, Axio HD can work with just about any type of file you can throw at it, which means intelligently handling all sorts of codecs, including native HDV. SD editing is well covered too, of course.

Axio HD is only available through approved dealers as a ready-built, turn-key system. In other words, you can't buy just the hardware to install in your own computer. The Matrox Axio HD *hardware* is priced at around $11,000 at the time of writing.

Current Digisuite owners can benefit from a substantial discount when upgrading to Axio, and if the system appeals but you are on a limited budget, you can start off with the SD version and upgrade later to the HD add-on.

Sony Vegas

Sony has worked hard to get its Vegas software noticed as a serious contender in the world of editing, and the uphill battle

Editing Hardware & Software

seems to be paying off. Vegas has earned the respect of many professional editors because of its unique and powerful approach to handling video, particularly its remarkable scaling abilities that were largely brought about due to the program's original development as an audio editor.

Image courtesy of Sony

If you're familiar with another editing program to the extent that you know it inside out, then there's probably not much point changing, but if you are fairly new to the world of editing then you might want to consider Sony Vegas. Like any new program it will take awhile to learn your way around the interface, but it's a capable

Like Adobe, Sony teamed up with CineForm to offer HDV support, with the option to upgrade. From version 6 onwards, Vegas also gives you the option of working with native MPEG files, but the sluggish responsiveness in that mode with anything other than a very fast computer makes it an impractical option.

Vegas also allows you to preview the timeline in DV resolution over the Firewire cable, which is of negligible benefit when you are working with HDV, but at least it's there if you need it.

Most of the transitions, filters and effects are available for use with HDV in real-time, although the preview output is not always as high resolution as with Adobe Premiere Pro.

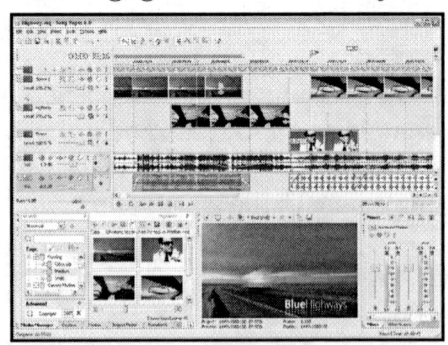
The interface and functionality of Vegas is winning over many editors
Image courtesy of Sony

102

How to Shoot, Edit and Distribute HDV

One of the best aspects of Vegas is its unique ability to scale video effortlessly, which means you can combine any number of different formats on the same timeline and work with them, regardless of your final output resolution.

Pinnacle Liquid Edition

Pinnacle, now owned by Avid, is a big player in the world of video. Some have theorized that Avid wanted to tap into Pinnacle's substantial market share of consumer-based editing products, in order to guide the more serious editors up towards the higher priced and better specified Avid software. Whatever the case, Pinnacle has continued to put out quality products geared towards consumers and semi-pro editors.

Like Canopus, their preference seems to be to produce their own dedicated editing software and hardware solution as an all-in-one package.

Compared with some of the other native HDV options, Liquid Edition manages to accomplish the process with less 'pain'. Yes, you need a very fast computer with plenty of RAM and fast hard drives, but if you meet those requirements you might be pleasantly surprised, especially if you have tried some of the other native editing offerings.

Of course, once you get into applying any kind of effects or filters, you can't hope for true real-time functionality, but when it comes time to export out to the HDV camera or deck, only footage that has been manipulated with effects, transitions etc will need to be rendered. The remaining footage is ready to go, which means that exporting is much faster than with solutions that transcode the video.

A downside to Liquid Edition is that currently it's not possible to monitor the video output on an external HD monitor—only one of the computer screens in a dual monitor setup can be used. With a high resolution computer screen you can get a good idea of what the project will look like, but it's preferable to have a permanent component HD monitor output.

Pinnacle Studio 10

The release of version 10 of Pinnacle's Studio software in October 2005 provided support for HDV for the first time. Also significant was the division of the Studio range into two subsets, 'Studio' and 'Studio Plus'.

The Studio package is designed for entry level consumers who want a simple and fast way to edit their videos and share them with friends. Studio Plus has more advanced functionality, including special effects and a wider choice of transitions, for the more experienced user or serious hobbyist.

Apple iMovie HD (part of iLife)

As part of Apple's inexpensive iLife package, iMovie is designed for beginners as a basic introduction to video editing. There is a lot of automatic-style decision making and hand-holding with the built-in 'wizards' such as 'Magic iMovie'—which may be perfect if you are new to editing. Good quality results can be obtained, and the software fully supports HDV. Projects can be output to DVD using the associated package iDVD, or exported back to the HDV camcorder.

Apple Final Cut Express HD

Final Cut Express HD is the slimmed down version of the Final Cut editing program, somewhat similar to Adobe's 'Premiere Elements' concept. Unlike the full version of Final Cut Pro HD however, 'Express' uses the transcoding method to edit.

Since the interface looks and feels very similar to Final Cut Pro HD, the upgrade path to the full version is made much easier than with most other software options. However, this also has a downside in that the interface does not appear to be beginner-friendly. If Final Cut Express is supposed to be an introductory product for enthusiasts, then the level of complication may put many users off HDV video editing.

Title software (LiveType) and royalty-free music creation software (Soundtrack) are also bundled with the software, making it excellent value for the Mac user who is serious about getting into HDV production but who has limited experience with editing.

One feature that is a thoughtful and sensible touch is the ability to import iMovie projects, which will be appreciated by those who start out on iMovie and then want to upgrade to something a little more powerful.

The software works with the OS X operating system and is relatively inexpensive, which makes it good value and within reach for most Apple users.

Apple Final Cut Pro HD

Despite many eager Mac users chomping at the bit for HDV support, Apple were another company who could have sold a lot more product had they brought it to market earlier. Ironically, in January of 2005, Steve Jobs—Apple's CEO—made a high-profile announcement declaring 2005 to be the year of High Definition, but then Apple left their users waiting for a considerable time to introduce full support for HDV in their flagship product, Final Cut Pro HD.

Built for the OS X operating system, Final Cut Pro HD is the software that Apple gears towards professional editors. It's

Editing Hardware & Software

designed to work with multiple video formats; in fact just about anything you can throw at it.

Final Cut Pro HD is the only Apple offering that edits native HDV. Both iMovie and Final Cut Express edit using an intermediary codec. Further, one of the software's big selling points is the ability to edit native HDV frame accurately. In other words, you are not limited by the usual GOP structure problems, any frame can be used as an edit point.

The 'Dynamic RT Extreme' feature provides intelligent playback of video, depending on the CPU power present in the system. Faster systems will allow for video playback with greater frame rates and quality.

Dual displays can be used with Final Cut Pro HD, which makes the editing process much easier.

Like Avid, Final Cut Pro has multi-cam editing functionality, which will be of great interest to you if you already use that kind of technology with DV. Numerous clips (viewing angles) can be viewed or cut at the same time.

One of the other major differences with Final Cut Pro HD compared with its sibling, Final Cut Express HD, is the inclusion of advanced three-way color correction tools and hardware accurate broadcast-style vectorscope and waveform monitors.

Final Cut Pro HD also features support for DVCPRO formats and HD video up to 10 bit uncompressed HD, so if you upgrade your equipment down the line or have a project where footage was shot with HD cameras, the software can cope with that.

Lumière HD

Lumière HD is the solution that many Mac users looked to before Final Cut fully supported HDV. Actually, some still use it because it uses the 'proxy' system of editing, rather than Final Cut Pro HD's native editing.

Lumière HD opens up the opportunity to capture and edit HDV even on a modestly powered Mac Powerbook. How? Lumière HD 'translates' clips to Final Cut Pro (or other editing software), using its XML interchange format. It's a simple matter of importing the XML file that Lumière HD generates into Final Cut Pro, and then you can start to edit, with real-time capabilities.

When it comes time to output your final project, the timeline is conformed to Lumière HD's QuickTime friendly, full resolution MPEG-2 HD. XML enables the re-linking of clips, and the whole process is very fast. Depending on which Mac you use, it is possible to view real-time HD output by using the DVCPRO HD codec. This is possible on a G5 for example. Other less powerful Mac's can display your output in standard definition.

Image courtesy of
Lumière HD

Chapter 19
Choosing & Using HDV Computers

Computers are great when they are working well, but if you've been around video editing for a while, you'll know that they can also spell trouble. In fact, they can take years off your life! HDV only complicates the situation because the format places unusually high demands on a computer system.

What's the Cost of an HDV Editing System?

One answer to that question could be another question—"How much can you spend?" Like so many other purchases that involve electronics, it depends on what your budget is. You could spend as little as $2000, or as much as $20,000 on a system that edits HDV. Obviously, the more you spend, the more speed, functionality, and real-time features you will gain. For example, a high-end system (such as the Matrox Axio) will allow you to edit HDV very easily, but you would also be paying for a lot of features that you might never use.

If your budget is at the lower end of the scale, you can put together a respectable HDV based editing computer for under $2,500:

>   **Dual processor computer for $1,650**
>   **PCI Express graphics card for $350**
>   **2 disk SATA RAID for $450**

Again, why is a heavy duty computer required to work with HDV? Remember, HDV's long-GOP MPEG-2 format is very CPU and disk-intensive. There's about four times as much data to decode and work with compared with DV.

Building Your Own Editing Computer

If you've put together your own editing system for DV in the past, you will have probably gone through the pain of discovering that choice of components is critical. In fact, lack of knowledge of this aspect is one of the most frequent causes of frustration and even despair!

Please take careful note of the following sentence, as it will save you many hours, days, or even weeks of heartache:

> **If you are not experienced with computer hardware, do not attempt to build your own HDV editing system!**

Ignore that advice and it will almost certainly all end in tears.

Building a stable, rock solid DV editing platform is difficult enough; HDV definitely adds to the complication. If you are *very* competent with building computers then you might feel inclined to build your own system, and if you do then there is one single piece of advice that is more important than anything else:

Check the manufacturer's website to see which hardware has been tested and approved as compatible.

Sounds simple, doesn't it? If only you knew how many people rush ahead to buy or build a computer, and subsequently try to install editing software or hardware that is incompatible. It happens constantly. Need some proof? Just check any of the software provider's forums online, and see how many people didn't use compatible hardware. It's a very expensive mistake...

Even when using recommended hardware, there are often little tweaks or installation procedures that have to be followed in order for the system to function properly. Pay very close attention to this information as it will save your sanity.

The last tip on this point is to regularly check for firmware updates for all your hardware and software. These occur fairly frequently. Get yourself on the email lists of the various providers of the components in your system. However, do use caution with updates. Before installing any update, do a little research on the forums to see what other users report. Are there any issues with the update? Does it solve some problems but bring about others? Ultimately, you might decide you are better off with your current setup, especially if your system is stable and functions well.

In summary then, only use hardware that has been tested and certified to work with the particular editing system that you want to use. This encompasses your choice of motherboard, graphics card, sound card, and hard drives; in fact all system components. Check for recommended installation procedures, and keep firmware and software up to date.

Intel or AMD?

The Intel or AMD debate for video editors seems to have been raging forever. Most experts on this subject agree that both companies have produced processors that have at one time been superior to that of the competing company, so it's constantly changing.

One thing you should be aware of is to take the advice of editing software and hardware providers with a large pinch of salt. For example, one well known manufacturer of video editing products has for some years promoted Intel as the best choice for performance and compatibility with their products. Little did most editors realise that this promotion seemed to have more to do with a marketing agreement than whether Intel was the best fit for their product. How interesting that in recent times this same company suddenly switched loyalties to AMD!

If you want unbiased advice on this subject, keep up to date with any of the respected hardware review sites on the Internet, such as www.tomshardware.com. Anyone who is in the industry but who is not trying to sell you a processor is probably a good source of advice.

AMD Dual-Core Opteron processors and Intel Dual-Core Xeon processors both perform very well. Going out on a limb, I would say that with AMD being more competitive on price, they have the edge...

Windows or Mac?

Er, let's not go there...

Power Supplies

One of the most frequent causes of failure or faults with video editing computers is the result of using a power supply that is not powerful enough for the job.

If you find that your computer fails to boot, stops part way through booting, or hangs (freezes), your PSU is likely to be the culprit. Think about it. Editing computers generally use more hard drives, more power-hungry graphics cards (especially with PCI-Express), and dual processors. All of those elements draw considerably more power than the average desktop computer.

Use at least a 500W PSU. In some cases you might need even more power. Yes, larger wattage PSU's are more expensive, but it's false economy to install anything less, as you will almost certainly end up having to upgrade it later anyway.

Choosing and Using HDV Computers

Graphics Cards

Before you buy a graphics card, ensure that your motherboard has a PCI-Express slot, as nearly all of the latest graphics cards use this fast architecture.

Don't skimp on your purchase of a graphics card, especially if you want to preview a full resolution HDV image on an HD monitor. Video editing is a specialist application and you need a specialist graphics card with enough power and resolution to cope. The chapter on monitoring HDV will give you some suggestions in this area.

Pre-built Editing Systems

The advantage of buying a turn-key (pre-built) editing system is that someone else has done all the hard work of developing a computer that is compatible and stable. You can just plug it in and start editing. For some people that's exactly what they want and need—they don't ever want to be hassled with having to open the computer cover.

Another advantage with buying a pre-built system is support. When something does go wrong, you can pick up the phone and receive knowledgeable help. For the non-technical, that can be invaluable.

Of course, there is a price to be paid for such convenience. System builders have to charge a reasonable fee to remain in business.

In summary, unless you enjoy building computers and are very good at it, you are better off buying a pre-built turnkey system. Even if you are on a tight budget, it's often false economy to try to build your own computer because when you factor having to replace incompatible components and the time spent trying to stabilize a faulty system, the costs can quickly overtake what it would have cost you to buy a pre-built system.

Incidentally, I highly recommend 'Guy Graphics' in Utah for pre-built systems. Their staff are very knowledgeable about editing equipment, they are up-to-date with what's available in the marketplace, and their prices are reasonable. Their website is: www.guygraphics.com. Please let Ashley Guy know that I sent you — he might even give you a better deal!

Detecting the Camera

If you are a Windows XP user, (for video editing you should not be using anything earlier than XP or 2000), ensure that you have SP2 installed (service pack 2) before connecting your camera via the Firewire cable to the computer. Windows should recognize the device automatically and correctly. Note that when you switch to the other mode within the camera (DV or HDV), Windows will see the connection as another device, which is normal.

Hard Disk Space & Configuration

When capturing HDV via Firewire, be prepared to consume about 40–60 GB an hour at 1080i. Obviously, if you convert the transport stream to an intermediary codec for editing, that extra space also needs to be taken into account.

What type of drive setup should be used? Although 7200RPM drives can *capture* HDV with ease, it's the *editing* stage that places the real demand on the drives. Editing HDV, regardless of the editing system used, is not just about reading data from the drive — the drive needs to *process* the data. If you intend to edit multiple layers of HDV footage, then a RAID is essential.

Choosing and Using HDV Computers

If you are not familiar with RAID, don't fret. It's an acronym that stands for *Redundant Array of Independent Disks* and it is a system that links two or more hard disks together for better performance. There are a number of different configurations, with a respective number for each one, and for the purposes of editing HDV the recommended type of RAID configuration is '0'. RAID 0 links drives so as to optimise the data transfer speed.

What type of hard drives should you buy? SATA drives are the preferred choice because they are faster than IDE, they are easy to add, and they are becoming the standard choice amongst editors. Be sure to check your motherboard accepts SATA drives, as some older motherboards do not. All current boards support SATA.

Also, remember that it's vital to use identical drives for a RAID setup, otherwise the RAID will not function properly.

Optimising Your Computer

As mentioned earlier, editing video, regardless of the format, places extreme demands on a computer, so you should tweak your system to be more efficient.

Programs that are set to run as soon as the computer boots up hog valuable system resources. If you have lots of little icons in the bottom right of your screen, then you are a prime candidate for a 'clear out.'

The good news is that there are a number of things you can do to prevent these 'blockages' that play havoc with video editing.

<u>Note:</u> If you are not fully competent with the way computers work, do not attempt the following procedures yourself. Ask a local computer tech guy to help you out. It's important!
You can find these problematic 'gremlins' in three places:

1. From the 'Start' menu, click on 'All Programs', then view the 'Startup' folder. Any non-essential programs in that folder should be deleted so as not to run at 'startup'. Do not attempt this unless you are sure of what you are doing.

2. From the 'Run' command, type 'msconfig' in the box to bring up the System Configuration Utility. Again, this is only for advanced computer uses. The 'Startup' tab allows you to identify and uncheck any items that you do not want running automatically in the background when the computer starts. Most computers have many items that can be unchecked to free up resources.

3. Some programs are set to run automatically at certain times of the day. For example, some anti-virus or anti-spyware software does this. Go into each program individually and either ensure that such updates are set to update manually, or set them to update at a time that won't affect your editing, such as in the middle of the night (your computer still needs to be on).

There are some other steps to take too:

1. Disable the desktop wallpaper (select a plain color background rather than an image or photo).
2. Disable the clock in the bottom right corner of your screen.
3. Disable screen-savers.
4. Disable power-management features that could turn off hard disks etc.

What else can you do?

Choosing and Using HDV Computers

Defrag Regularly

Regularly defragmenting ALL your hard drives is essential. Preferably, this should be done once a week, or after capturing. You would be astonished at how many times a sluggish editing project is due to the video files being fragmented. This is because video files are often very large, and when they are captured, the disk is working so hard to keep up with the data rate that it puts the information in the first available place on the drive. This often means it is spread across different parts of the drive—that's why they are referred to as being fragmented.

Although Windows has a defrag utility built-in, there are other faster, superior utilities out there that you should consider.

Multiple Drives

You need to use at least three hard drives in an editing system: one drive for your operating system, another to capture video (but preferably a RAID setup), and finally a drive to export your finished projects for storage. It's asking too much for a system to run the editing software AND capture to the same drive. You might get away with it with some DV systems, but you won't with HDV.

Games are Evil

Games are the evil enemy of editing computers. Never load games on your editing system, as they WILL hog an enormous amount of resources and they WILL play havoc with various behind-the-scenes settings to really screw your machine up when you come to edit. If you're into games, get another computer just for that! You've been warned!

Viruses and Spyware

If games are the evil enemy of editing computers then viruses and spyware are in a league of their own at causing havoc and leaving a path of vicious destruction. Sadly, these days it is unbelievably easy to be a victim of spyware and viruses. Some of the geekiest computer nerds that I know have been victims of serious viruses that destroyed the data on their computers. One survey showed that it was not uncommon for a computer that was connected to the Internet to be infected within two minutes of an operating system being installed!

What can you do about this major issue? Firstly, don't use email on your editing computer. It's just not worth it. No single anti-virus software is 100% effective, so why take the chance that one day you might lose a project that you had spent days, even weeks working on? Use a separate computer for all tasks other than editing, whether it is games, e-mail, word processing, accounts, or whatever. Don't think that you will be one of the 'lucky ones'. If you take no precautions at some point you WILL get a virus and regret it.

Secondly, install anti-spyware and anti-virus software and run it regularly. An excellent anti-virus software that many pros use is 'AVG' from www.grisoft.com. Microsoft offer a utility to detect and destroy spyware, called Windows Anti-Spyware. It can be downloaded from www.microsoft.com/downloads. Other options are also available.

Spend the Time and Reap the Rewards

All of these tweaks and adjustments might sound like a lot of hassle, but you will find that they make a *big* difference to the responsiveness of your system. The good news is that most of these changes can be made once and then left alone, so it's not as complicated or time-consuming as you might think.

Chapter 20
Useful HDV Devices & Accessories

As the world of HDV becomes more established, more and more manufacturers are coming up with useful add-ons, accessories, and devices to expand the usability and functionality of the format. This chapter considers some of these items that are available to you.

HDV-Connect by Convergent Design

HDV-Connect is a small, half-rack size standalone box that handles broadcast quality conversions to and from various signal types that you might come across when working with HDV. It supports 720p, 1080i, and SD inputs and outputs, along with HD/SD SDI, Firewire, DVI, and component connections.

The most obvious use for the box is to convert an HDV signal to HD SDI, DVI, or Component. For example, many high-end HD monitors work with SDI inputs so the HDV-Connect box would provide the interface that would allow you to output from your editing computer to the HD monitor.

It's very well designed, with many thoughtful touches, including the facility to update the unit's firmware over the Internet, using a standard LAN port.

Serious Magic Ultra 2

Accurate chroma-keying is one of the most difficult and painstaking tasks to accomplish with video. Fortunately, Serious Magic is an innovative company that has produced a piece of software that makes the process a lot easier, especially with HDV.

How to Shoot, Edit and Distribute HDV

Ultra 2, designed for the PC platform, supports both DV and HDV, including 720p and 1080i resolutions, at frame rates of 24p, 50i, and 60i. In fact, since Ultra 2 uses its own proprietary system of 'vector keying', you can work with any resolution.

Image courtesy of Serious Magic

One of the other major benefits to Ultra 2 is that keying is far simpler and more accurate compared with most built-in key capabilities of non-linear editing software. Even a slightly crumpled or unevenly lit green or blue screen seems to present no problem to the software; even a subject's shadows can be preserved.

Another feature is the support for high-resolution stills up to 4096 x 4096. This means you can use digital images as High Definition backgrounds. Once you have keyed them to your liking, you can even export a high resolution still image of the composition for use in web or print work.

One of the major drawbacks to most keying software is that you can't see how the shot will look in the final composition while you are still shooting. You might finish the shoot and only then find that the lighting does not look good, or that the angle you filmed from could have been better adjusted to suit the background. By installing Ultra on a laptop or computer near your green or

Chroma-keying is easy with Ultra 2

Useful HDV Devices and Accessories

blue screen, and connecting your HDV camcorder via the Firewire port, Ultra allows you to monitor the composition 'live', allowing you to adjust your lighting and composition to perfectly match the scene.

You can download a free trial of the software from the Serious Magic website: www.seriousmagic.com

Sony HVR-M10 HDV Player/Recorder Deck

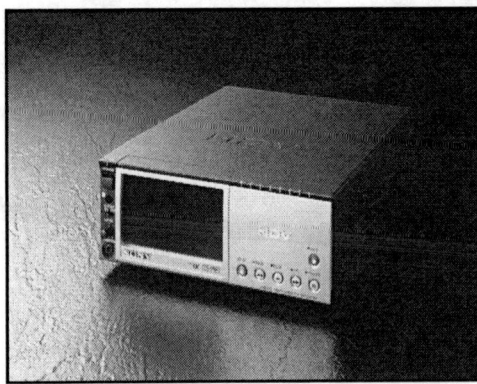

Considering it's their first HDV deck, Sony did a remarkable job. It handles all the functions you might need, frees up your HDV camera to be used for further shooting, saves wear and tear on the camera's heads, and even includes an LCD monitor on the front.

Image courtesy of Sony Electronics Inc

A thoughtful and logical touch is the way the M10 uses the same batteries as the FX1 and Z1 cameras. That makes it easier to use as a portable deck, without having to stock yet another type of battery.

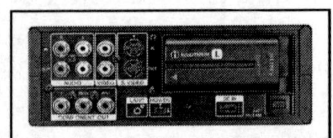

The rear of the M10 deck, with a battery attached

Connection options are audio in/out, Firewire in/out, composite in/out, S-Video in/out, and component out. The lack of a component input and the inability to record timecode are annoyances, but you have to consider that, at this price point, it's probably asking too much.

JVC BR-HD50 ProHD Player/Recorder Deck

Designed to be a natural pair with their HD100 camcorder, the JVC deck has some commonalities with its Sony counterpart, but as you might expect there are some major differences too.

Like the HD100 camcorder, the deck is also confusingly referred to as 'ProHD,' even though it conforms to the HDV standards—so consider it an HDV deck. Marketing people are sometimes too clever...

Image courtesy of JVC

The most obvious difference on the front of the deck is the lack of widescreen LCD display that the Sony sports. Instead, in the middle of the fascia is a large timecode display, along with audio meters. Not having an LCD screen to monitor the image is a disappointment. Another area where the deck falls short compared to the Sony M10 is that no battery option exists, which seriously limits its use to the studio environment.

Useful HDV Devices and Accessories

On the plus side, there are a number of features that provide an edge compared to the Sony, depending on what your needs are.

A definite plus point with the JVC deck is its ability to handle both sizes of DV and DVCAM cassettes, and all associated formats / frame rates can be played back with ease. Also on the positive side is the inclusion of an HDMI output port. HDMI is a digital signal connector that is increasingly being used on High Definition projectors, plasmas, and LCD screens.

RedRock Micro Adaptor

One of the key characteristics of the 'film look' is shallow depth of field. This is where the foreground subject is sharply in focus but the background is out of focus, to the point of almost being blurred. Cameras that use fixed lenses such as the Sony FX1 and Z1 make it difficult to achieve shallow depth of field. Recognising the issue early on, a company named RedRock Micro designed an adaptor that mounts on the front of the existing lens to allow you to use any 35mm lens. This in turn makes it relatively easy to achieve the appealing look of shallow depth of field.

Image courtesy of RedRock Micro

Actually, the use of 35mm lenses opens up a lot of creative possibilities, with varying angles, improved zoom capabilities, and so on. Mounts are available for Canon MF, AF, Nikon PL and OCT-19, and Pentax; even inexpensive 35mm SLR lenses can be used. If you are a film student or seriously interested in imitating the look of film by using techniques such as rack focus and follow focus with film-like lenses, this product will likely be of interest to you. Amazingly, it's priced extremely reasonably (under $600) for a relatively complex device that delivers so much.

Chapter 21
Monitoring HDV

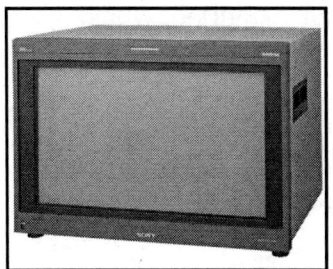

As mentioned earlier, the ability to accurately monitor full or near-full resolution HDV while recording and editing is essential. The challenge is that full resolution HD monitors are extremely expensive. Unbelievably expensive in fact. Admittedly, a few companies such as Marshall have worked hard to bring prices down, so you may want to look at their slightly more affordable HD monitors.

COLORSPACE

Something important to note on the subject of monitoring is that **HDV has a colorspace of 709.** This means that CRT monitors connected to an HDV source will not display accurate colors, nor in fact will any monitor that is not designated as HD or HD ready.

LCD or Plasma? Which is Best?

For video editing, plasma has the edge because LCD screens tend to have viewing angle issues and don't display true blacks as well as plasma screens. On the other hand, bear in mind that plasmas have an issue with burn-in. You can't keep a still image on a plasma screen for any length of time because it will burn into the phosphors and leave a permanent impression. One other little known issue with plasmas is that they won't work at high altitude, so that may be a factor for you to consider, depending on where you live or work.

Monitoring HDV

Price may also be an issue for you. LCD screens in the 20"–28" are usually substantially cheaper than plasma screens. For large screen sizes above 32", a plasma will usually offer the best value.

When buying a monitor, be sure to note the resolution, which is the single most important factor in the context of editing HDV. Not all monitors are created equal, and an LCD screen of 20" could have a resolution as low as 800x600, or as high as 1600x1200. Clearly, the former would not be acceptable for monitoring HDV, whereas the latter would be ideal.

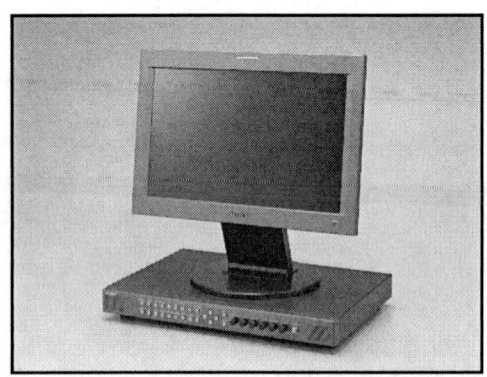

Sony Luma LCD monitors are specifically designed for HD
Image courtesy of Sony Electronics Inc

Also, be aware that very few monitors currently display full HD resolution. That applies even to large plasma screens. Apple Cinema displays are one of the few choices that can display full resolution HD, so they are a popular choice, but they do have a limiting factor in that they only support a DVI input.

As a guide, your monitor should be able to display 1080x768 at the very least. At that resolution you will see a noticeable improvement compared with standard definition, but for accurate monitoring you really need a monitor that can display 1280x1080 pixels or higher.

These rack mounted monitors can be used to monitor multiple sources in the studio or field
Image courtesy of Sony Electronics Inc

How to Shoot, Edit and Distribute HDV

> **Suitable HD Monitor Options Include:**
>
> Dell 2405FPW (www.dell.com)
> Westinghouse Digital LVM-37w1 (www.westinghousedigital.com)
> Apple Cinema Displays (www.apple.com)
> HP L2335 or F2105 (www.hp.com)
> Samsung 730MW or 242MP (www.samsung.com)
> Sony SDM-S204 or SDM-P234 (www.sony.com)

Graphics Cards with Component Output

In the very early days of HDV, one of the most frustrating things was the inability to preview a full resolution HDV output from the timeline onto a full resolution display. Thankfully, several graphics card manufacturers came to the rescue, and it's now possible to see full resolution HDV on an external monitor while editing.

The introduction of the PCI-Express architecture probably played a part in the availability of these cards, because they support an unprecedented level of data transfer bandwidth. Note that PCI-Express should not be confused with PCI-X, they are not the same and they are not compatible. If you are in the market for a new motherboard for video editing and you want to use the latest generation of graphics cards, you must specify PCI-Express.

n-Vidia Quadro FX540 Graphics Card

Although n-Vidia are probably best known for their products in the gaming field, they have been gaining increased recognition in recent years for their high-end graphics cards for specialist applications, especially for the financial and CAD industries.

The NVIDIA Quadro FX 540 graphics card features HD Component, S-Video, and composite video connections, all

Monitoring HDV

accessed through an external breakout box. The card has 128Mb on board memory and uses the PCI-Express architecture. It works with major video editing applications that support video overlay and the Open GL architecture, such as Pinnacle Edition and Avid Xpress Pro.

The main feature of this card is the ability to spread your editing application across two desktop monitors, with simultaneous component video output up to the full resolution of 1920x1080i (depending on what your monitor can display). However, the downside is that the monitor outputs consist of just one DVI output—the other is an analog VGA connection.

Matrox Parhelia APVe Graphics Card

Image courtesy of Matrox

The entire Matrox Parhelia range is well regarded by video editors and at the time of writing, the 128Mb Parhelia APVe had the distinction of being the only graphics card that had two DVI computer displays plus a simultaneous video output. Although you can't output to two *DVI* monitors with an HD monitor simultaneously, you can have just about any other combination of connections. This means it is possible to work on dual computer monitors (one analog and one digital), while *also* previewing HDV

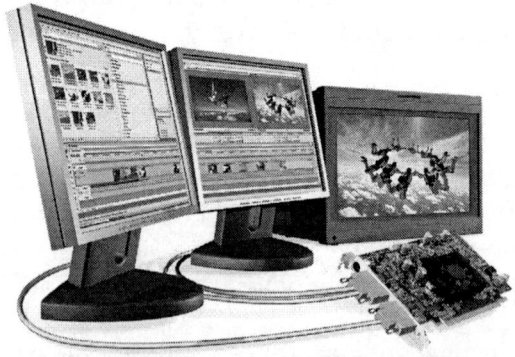

The Matrox APVe allows you to spread your timeline across two computer monitors and also output full 1080i resolution on an HDTV or third computer LCD screen
Image courtesy of Matrox

output on an external HD-capable monitor via component or VGA connections. The dual-display output is available up to an amazing 1920 x 1440 per display.

In case you were wondering, APV stands for Audio, Photo, and Video. So how does it measure up? Frankly, this card is the answer to an HDV editor's prayers! To have an editing timeline spread across two monitors *and* see a real-time 720p or 1080i preview on an HDTV monitor might seem like a luxury to some, but it really is a basic requirement for working with HDV that was unavailable until recently.

The combination of displays that can be used is truly impressive, and all the possibilities (DVI, D-SUB (VGA), S-Video, Composite video and component video) are easily connected by means of an assortment of cables that come in the box. The card uses the lightning fast x16 PCI-Express architecture found on most modern motherboards, and works seamlessly with Adobe Premiere Pro, Adobe After Effects, Sony Vegas, Ulead Media Studio Pro, and Avid Xpress Pro.

The supplied 'Power Desk' driver software is very stable, and easy to use. It includes the ability to rotate the image for display in portrait or landscape (pivot mode), duplicate a single desktop image across multiple displays (clone mode), or magnify a section of the main screen to be viewed on other screens (zoom mode). One feature I would have liked to have seen is the ability to toggle between SD and HD TV outputs. Maybe in a future update?

Graphics cards are traditionally notorious for being the weak link in the stability of a computer system. They seem to have more conflicts with more software than any other part of the hardware. Fortunately, this genuinely does not seem to be the case with the Matrox APVe. It is rock solid and reliable. Bravo, Matrox!

Chapter 22
How to Distribute HDV

One of the most frequently heard complaints about HDV has been the lack of distribution methods available for the format. Until Blu-ray and HD-DVD become widespread, you won't be able to just "burn a High Definition DVD" and play it instantly.

Although this distribution challenge has dissuaded many from even entering the world of HDV, the real question that you should start out asking yourself is, "How do I distribute my DV projects at the moment?" The answer to that question will help you to decide whether HDV distribution will be an issue for you.

Let's explain this...

If your main business is producing DVD's for retail sale or for 'end use' by a consumer, you will need to wait until HD-DVD and Blu-ray become widespread. However, if you produce any kind of video for purposes such as training, exhibition display, promotion, corporate, or for high-end consumers, there are options available right now that will enable you to distribute your project relatively easily and inexpensively.

Are DVD's High Definition?

Let's clear up a misunderstanding before we delve deeper into this topic: standard DVD's (DVD-R, DVD+R and the re-writeable formats) are *not* High Definition. The reason why this seems to be an area of confusion for many people is because the picture quality of DVD is so much better than the analog VHS tapes that were the predominant format for so many years.

Having said the above, bear in mind that DVD stands for 'Digital Video Disc' or 'Digital Versatile Disc'. That should remind us that

DVD's *now* come in a variety of formats—Blu-ray and HD-DVD are also referred to as DVD technology, but these new variations of DVD can play true HD and HDV. Since this technology is still in its infancy, for all intents and purposes it is correct to say that DVD is not HD, because at the time of writing, most people still think of DVD in relation to the movies they can rent or buy.

Currently, many programs that were shot in HDV end up being distributed as SD DVD's (standard definition). Obviously that will change as distribution methods for HDV become more standardised and widespread.

To remind you, DVD's that are created from HDV footage are better quality than DVD's originating from standard definition video. Why is that? Quite simply, HDV provides more pixel data to work with to begin with, so there is less reduction in quality in the MPEG-2 transcoding process. Try it for yourself. Standard definition DVD's produced from an HDV project look stunning.

Distribution Options

Getting back to the key point of this chapter, contrary to popular belief it's not all doom and gloom when it comes to distributing and viewing your HDV projects. There are some distribution options available to you right now, so let's discuss the pros and cons of each of these.

WMV-HD

Microsoft is heavily committed to HD, and their contribution to its distribution in the form of the WMV-HD format is significant. WMV-HD, which stands for Windows Media Video High Definition (now you know why they shortened it!), uses a clever compression algorithm to squeeze the enormous standard file size of a finished HDV project into a much more manageable file size.

How to Distribute HDV

In fact, it is possible to fit an average length HD feature film on a standard recordable DVD.

As well as the ability to play back on a computer equipped with a powerful processor and high-end graphics card, there are two companies that offer DVD players that play WMV-HD files: Kiss Technology (www.kiss-technology.com) and I-O Data (www.iodata.com). The data is burned onto regular DVD's in a similar fashion to creating standard DVD's. I-O Data's unit is called AVeL Link Player, and Kiss has multiple models available.

WMV-HD is a practical method of distributing HDV projects until HD-DVD and Blu-ray technology becomes widespread and cheaper. For example, for less than $250 to buy a WMV-HD capable player, you can provide a client with the ability to easily replay their project in full HD resolution.

The only downside to this solution is that currently, few options exist to create conventional DVD style menus, so the on-screen menus of WMV-HD discs are basic and functional.

DIVX-HD

DivX-HD is a compression format similar to WMV-HD, but it is not as widely used. It compresses HDV down to a size that is small enough to be sent over the Internet. DivX HD files can be up to five times smaller than regular HD files.

Currently, the technology only supports up to 720p resolution at 4Mbps. 1080i is not supported at the time of writing.

A full length DivX HD feature can fit onto a DVD, so that's good news, but you will need a DivX HD compatible DVD player. Currently, I-O Data, Kiss Technology, and a company called Buffalo Tech (www.buffalotech.com) produce compatible players for DIVX HD.

How to Shoot, Edit and Distribute HDV

> **Don't Forget 24fps**
> If you decide to distribute a standard definition version of your project, bear in mind that 24p DVD is an option for you, especially if your project was filmed in 25fps or 24p. As a reminder, 24p will give you more of a 'film look'. It also has the advantage that, since you are losing around 20% of the frames compared with 30fps, the overall project will be smaller, allowing you to use a higher bit-rate (less compression).

D-VHS

Digital VHS technology from JVC has been around for longer than HDV—since 2001 in fact. Ironically, for a short while before HDV came on the scene, it seemed to have virtually faded into obscurity. However, as a result of HDV it has seen a great resurgence of interest.

As the name suggests, D-VHS is a tape format that uses the same size tape as VHS, but records the images digitally rather than by the traditional analog methods employed by VHS.

D-VHS decks are made by JVC and Mitsubishi, and bearing in mind that their average selling price was around $2000 on their release, some models can now be purchased for as little as $300, so they are relatively inexpensive. Note that most models only accept Firewire input, not component in, which is a limitation.

HD1 (720p) formats can be transferred onto DVHS tape via Firewire easily, but the standard bit rate of HD2 is too high for D-VHS. The solution is to lower the data rate to below 19Mb/s, to make the HD2 stream compatible. Most editing software will allow you to do this by exporting (re-encoding) your project using a D-VHS template. If there is no template, look for a way within your software to create a 1080i transport stream with a bit-rate of around 14Mb/s. You can then export the transport stream to the

D-VHS deck in the same way you would export it back to the camera.

Given the fairly low cost of D-VHS recorders and tapes, it may be a suitable tool for storing your completed projects. Obviously, one of the downsides is having to re-encode 1080i footage to make it compatible, a process which obviously degrades some of the quality. The other downside is that it's a tape based format, which means it is more difficult to navigate (you have to rewind and fast forward the tape), plus it is more prone to degradation while in storage.

DVD for Archiving

If you find yourself working on a large number of HDV projects, DVD may be a suitable backup / archive tool for you. Depending on the length of your project, most transport streams will fit onto a standard recordable DVD or a dual-layer DVD. It's a digital copy of the data, so you won't lose any quality in the process, but for precious projects it's worth backing up on another format in addition, perhaps a hard disk. I would not recommend using a DVD or hard disc exclusively for backup as hard disks can fail and DVD's can get scratched, causing difficulty in accessing the data later on.

HD-DVD & Blu-Ray

These two formats represent the next generation of recordable DVD technology. With far increased storage capacities of up to 50GB, this is definitely the way forward. This technology will enable the widespread distribution of HD content without having to resort to heavy compression.

The problem is that HD-DVD and Blu-Ray are competing formats, much like the famous Betamax / VHS war, and more recently the competing +R / -R recordable DVD formats. Taking the latter example, with current recordable DVD technology there is now something of a stalemate, with many recorders able to record and playback both −R and +R discs. However, when it comes to choosing an HD consumer playback format, there can be only one winner. Why? The technologies to record and play the two formats are radically different from each other, so it's unlikely that any compromise will be reached. One format will win out over the other.

HD-DVD and Blu-Ray, which both support the H.264 codec, have both been promised and eagerly anticipated for an extraordinary length of time now. 2005 was publicly touted as the year for these formats to become available, but the year came and went without the recorders making their entrance. No previous format seems to have received such a fanfare of publicity for such a long length of time prior to its introduction. It appears that numerous delays have created setbacks in launching the technology.

Even though 2006 saw their official launch, widespread adoption of the formats will still take some years because of their high purchase price and teething problems.

As you might expect, each of the two formats has advantages and disadvantages. Let's look at each format in detail.

HD-DVD

Supported by Sanyo, Toshiba & NEC, as well as the DVD forum (an organization that develops and defines DVD formats), HD-DVD features a storage capacity of up to 30 GB (dual-layer), compared with 4.7 GB for the current, standard single layer DVD. According to Toshiba, a 45GB triple layer disc is in development, as well as discs that feature standard DVD style content on one side, and High Definition DVD content on the flip side.

How to Distribute HDV

Since HD-DVD discs have some commonalities with existing DVD's, they can be manufactured with much of the same equipment and will therefore be significantly cheaper to produce compared to Blu-Ray discs. For consumers, this compatibility of HD-DVD with older discs means that they will be able to play all their DVD's—both old and new, in an HD-DVD deck. In other words, the players will be backwards compatible.

Clearly, the reduced production costs *and* compatibility with existing DVD's gives HD-DVD a potential advantage in the marketplace.

Blu-Ray

Supported by Sony, Panasonic, Philips, and Apple, Blu-ray uses a slightly thicker disc than regular DVD's, and a blue-violet laser reads the disc, hence the name Blu-Ray.

The disc can store up to 50 GB of data (dual-layer), which is a substantial competitive advantage over HD-DVD.

However, Blu-Ray has a couple of major negative factors weighing against it. First and foremost, the discs are not backwards compatible with current DVD discs, so at the time of writing it's not clear whether Blu-Ray DVD players will be able to play current DVD's. It's possible that manufacturers could add another 'read' head but that will certainly add to the production costs of making these units.

Another factor weighing against Blu-Ray is that the protective surface layer of Blu-Ray discs is just 0.1mm, compared with the 0.6mm of standard DVD's. This makes the discs expensive to produce and more susceptible to read problems when the discs are not handled properly.

In terms of marketing, Blu-Ray has scored a significant early win in that Sony have promised to use Blu-Ray drives in their next generation of gaming consoles, which will probably be PlayStation 3. Getting Blu-Ray players into millions of homes early on through this method could give Blu-Ray a head start in the format race.

Will it be Betamax vs VHS Again?

Let's hope that the HD distribution formats do not end up in a 'war' like Betamax vs VHS

The situation with these two competing High Definition formats brings to mind the well-known Betamax versus VHS 'war'.

What complicates the situation is that movie studios have started to take sides with these two HD formats, and their market influence is another factor in the mix that could affect the outcome of which format becomes dominant. For example, Warner Brothers, Universal, and Paramount are backing HD-DVD, while Disney, Sony Pictures, and Fox are backing Blu-Ray.

As with the VHS / Betamax fiasco, the best specified format may not be the one that wins out. It may come down to the functionality and effectiveness of its copy protection scheme. This is an issue that is at the top of the agenda of every movie studio. They want to avoid a repeat of the poor protection of current DVD technology, which is relatively easy to copy.
Then again, the success or failure of one format over another may also come down to which format is marketed better. That was the case with VHS vs Betamax. Ultimately, only time will tell which format will win out...

135

How to Distribute HDV

Other Developments

Seemingly fed up with waiting for Hollywood and the major manufacturers to get their act together with BluRay and HD-DVD, in 2004 many Chinese electronics companies took it upon themselves to come up with a method for distributing HD content. Their solution, known as HVD, stands for *'High Definition Versatile Disc' or 'High Clarity Video Disc,'* depending on who you ask. The system uses a red laser and seems to be based around both the MPEG-2 and MPEG-4 codec's, using a chip manufactured by an American company called Amlogic Inc. Players such as the NeuNeo HVD2085 can be purchased inexpensively from www.neodigits.com. Another player, the Skyworth HVD-3050, even comes bundled with several HVD movie titles.

Then there's Taiwan. Not to be outdone by China, no less than 28 Taiwanese manufacturers have come up with yet another format which they call **FVD**, which stands for *Forward Versatile Disc*. It's another type of physical disc the same size as current DVD's, with capacities of 5.4GB as standard, or 9.8GB in the dual-layer structure. In this instance, the data storage size is not as critical as other formats because it uses the WMV-9 codec, a very efficient compressor that enables an FVD disc to easily hold 140 minutes of HD video.

The HVD format explained earlier is not to be confused with *another* future format bearing the same initials. **Holographic Versatile Disc** (also confusingly known as HVD) by Optware is a green laser format still in development, but promises to store around 200GB of data initially, moving up to a maximum of 1 terabyte. That's an incredible amount of data, so with news like that it will be very interesting to see the shape of things to come in other areas of the world of High Definition.

Chapter 23
Conclusion

Ever since HDV was released, the format seems to have divided videographers, with some strongly supporting it and others condemning it. So it's been a rough and controversial road already, and it's still early days for the format. One might reasonably ask, "Is HDV here to stay?" The answer in my opinion is, most definitely!

Is now the right time for you to get involved with HDV? In all fairness, it depends on what you plan to do with the technology.

As you've seen in this book, working with HDV is more challenging than DV in a number of ways. It is also more expensive, so those are two key considerations for you.

On the positive side of things, HDV provides the ultimate image quality within a reasonable price range. Also, it is the way of the future and you can expect it to overtake DV not just as a preferred choice, but ultimately as the new standard.

If you currently shoot video for a living, don't be afraid to dive into HDV. It's far better to go through the learning curve now rather than wait until clients are demanding their projects in HD while you are left floundering still learning the ropes.

Bear in mind that even now, many of your clients are seeing the merits of HD simply by watching their favorite TV shows in HD on their home plasma or large LCD screen. The knock-on effect of that is, clients are increasingly prepared to pay a premium for HD in their corporate projects. For example, they like the idea of displaying their high-def corporate video on a plasma screen at their next exhibition; they get excited about wowing audiences

Conclusion

with HD content projected onto large screens at a conference, road-show, or seminar. They can capture the process of how to manufacture a tiny widget (filmed in amazing detail like never before), and then train staff on how to repeat that process.

Wedding Videos are Prime Candidates for HD

Brides (and grooms) are starting to realize that their wedding only happens once (hopefully) and that even though they or their guests may not be able to experience the HD version of their video right now, it's likely that a year or two from now HD TV's will start to become the norm, and five or ten years down the line when their kids come along, they will definitely regret not having an HD wedding video. So weddings are a prime target for High Definition shooting opportunities.

Much more needs to be done to educate engaged couples and their families about the merits of HD, but the efforts of organizations such as WEVA are finally starting to pay off, and things are heading in the right direction. Such efforts should be encouraged, of course. The more people that are aware of HD, the better it will be for all of us.

In a broadcast environment, HDV is now finding a place in a variety of situations.

Firstly, HDV presents a practical solution for news crews working in dangerous environments or war-torn countries. Crews seem happier to use the current crop of lighter, ultra-portable HDV cameras, and the bean-counters in the corporate office obviously prefer to have a $5k loss than a $65k loss, if it came down to that.

Additionally, HDV cameras are increasingly being used in shooting locations where large HD cameras are impractical or too time consuming to set-up. This would include confined spaces and in situations where a reporter, director, or producer does not have a professional cameraman on hand but needs to acquire some high quality footage on demand.

A third area where HDV cameras have found a home in broadcast are in reality-based shows, documentaries, and long-term shooting environments where HD cameras would be prohibitively expensive in relation to the programme budget.

As you can see, in certain situations, rather than being an inferior sibling, HDV has some distinct advantages over large, expensive HD cameras.

Don't Wait for the 'Next Big Thing'!

If you are not making an income from HDV and have been holding back on buying an HDV camcorder because you are waiting for the next great model to be released, you are losing out now! There will always be a better model ahead, that's the way technology works. It's the same way with computers. Virtually every month a superior, faster, better computer processor comes on to the market. You have to draw the line somewhere and bite the bullet.

If you are putting off buying now because of comments by some that HDV technology has not settled down yet, the same could be said of DV. DV cameras are still being improved even now, and editing facilities definitely still have much room for improvement. If you wait, you will probably be left behind because HDV has already become main-stream and it is here to stay.

Try Before You Buy

Depending on where you live, opportunities may exist for you to rent an HDV camcorder from an audio-visual company for a day

Conclusion

or two. That's a great way to try out the technology and see how you get on with it.

The future is an exciting time for HDV. As HD TV's become cheaper and higher quality, they will rapidly become the standard choice amongst consumers. Along with that growth will be an increased demand for HD content. No doubt HDV will play an important role in acquiring and providing some of that content.

As computer processors get faster and manufacturers come up with more efficient ways of editing HDV, we can expect the format to replace DV. Over time we'll become so used to the quality of High Definition that we'll eventually look back with fondness at the 'quaintness' of 'old fashioned' DV!

Have fun with HDV!

One more thing. Your perseverance in reading to the end of this book has paid off! As per the promise on the front cover, you can now benefit from free updates to this book by sending an email to:

updates@hdvguides.com, stating where you purchased the book.

Additionally, I write a periodic HDV newsletter with tips, techniques, money saving offers and lots of other good stuff that you can't afford to miss. If you would like to benefit from this newsletter, send us an email to:

news@hdvguides.com with 'subscribe' in the subject line, and we'll keep you informed, for free!

Appendix:
HDV Related Websites & Resources

Mini-DV Tapes Suppliers
www.edgewisemedia.com
www.tapestockonline.com

Cameras and Equipment Suppliers
www.bhphotovideo.com
www.edgewisemedia.com
www.symbiosiseu.com

Add-on Lens Suppliers for Sony FX1 / Z1
www.centuryoptics.com
www.16x9inc.com
www.optexint.com

Software Plug-ins
www.redgiantsoftware.com
www.digitalfilmtools.com
www.dvfilm.com
www.nattress.com
www.thefoundry.co.uk

Training Products
www.hdvguides.com
www.thehdvstore.com

Direct to Disk Manufacturers:
Focus Enhancements (Firestore): www.focusinfo.com
Shining Technology (Citidisk): www.shining.com

Camera Manufacturers:
www.sonystyle.com or http://bssc.sel.sony.com
www.canon.com
www.jvc.com/pro

HDV Related Websites & Resources

Editing Software and Hardware Manufacturers:
www.adobe.com
www.apple.com
www.avid.com
www.CineForm.com
www.canopus.com
www.mainconcept.com
www.matrox.com
www.pinnaclesys.com
www.sonymediasoftware.com
www.ulead.com

Floating Camera Devices:
www.steadicam.com
www.glidecam.com

General HDV Information:
www.sonyhdvinfo.com
http://adamwilt.com
http://hdsource.highlydef.com
www.hdvcafe.com
www.hdv-info.org
www.hdvinfo.net
www.highdefforum.com

Forums:
www.hdvforum.net
www.dvinfo.net
www.wwug.com
www.creativecow.net

Do You Own a Sony HDV Camera Or are You Thinking of Owning One?
These Instructional DVD's are a 'must have'...

DVD for the FX1 and Z1 Sony HDV Camcorders

HandsOnHDV:
"A Complete Guide to the Z1U & FX1 Camcorders" was shot entirely in HDV with Z1 and FX1 camcorders, edited with FCP, and output in 16x9 SD for the DVD. Throughout the video, a variety of video shooting modes are demonstrated and explained, including Cineframe 24, CinemaTone, and Picture Profiles.

This is not a bench-top demonstration or in-studio camera review. It is an in-depth training video with shooting done by professionals with years of ENG/EFP broadcast shooting experience.

An **EXTRAS** folder is included on the DVD with several exclusive documents, including: A detailed reference chart that outlines every customizable setting of all six of the default Picture Profiles so you can compare them and see the differences. Also included is a comparison chart of the major differences between the Z1 and the FX1. Some features discussed in the video only apply to the Z1 and this chart helps identify those differences.

DVD for the Sony HC1 HDV Camcorder

HandsOnHDV:
"How to shoot like a Pro with the Sony HDR-HC1 Camcorder" was created for anyone who isn't satisfied with just picking up an HC1 and shooting in fully-automatic mode. It is a complete camcorder instructional video and an in-depth guide to advanced shooting techniques all in one package.

This video goes well beyond the owner's manual in every way—plus it's much easier to understand—two very important things to consider when tackling a production tool as complex as the HDR-HC1.

If you want to learn how to get professional results with the Sony HDR-HC1 High Definition camcorder—you need this DVD. It's packed with two hours of tutorials, examples, and detailed instructions to help beginners and advanced videographers of all skill levels quickly master this amazing little camcorder.

For more information, or to grab your copy of either of these excellent DVD's, visit: www.hdvguides.com or call 1-888-707-7595